T0220239

Cambridge Elements ≡

Elements in the Philosophy of Science
edited by
Jacob Stegenga
University of Cambridge

FUNDAMENTALITY AND GROUNDING

Kerry McKenzie
University of California, San Diego

CAMBRIDGE
UNIVERSITY PRESS

CAMBRIDGE
UNIVERSITY PRESS

University Printing House, Cambridge CB2 8BS, United Kingdom

One Liberty Plaza, 20th Floor, New York, NY 10006, USA

477 Williamstown Road, Port Melbourne, VIC 3207, Australia

314–321, 3rd Floor, Plot 3, Splendor Forum, Jasola District Centre,
New Delhi – 110025, India

103 Penang Road, #05–06/07, Visioncrest Commercial, Singapore 238467

Cambridge University Press is part of the University of Cambridge.

It furthers the University's mission by disseminating knowledge in the pursuit of
education, learning, and research at the highest international levels of excellence.

www.cambridge.org
Information on this title: www.cambridge.org/9781108714020
DOI: 10.1017/9781108657617

© Kerry McKenzie 2022

First published 2022

A catalogue record for this publication is available from the British Library.

ISBN 978-1-108-71402-0 Paperback
ISSN 2517-7273 (online)
ISSN 2517-7265 (print)

Fundamentality and Grounding

Elements in the Philosophy of Science

DOI: 10.1017/9781108657617
First published online: February 2022

Kerry McKenzie
University of California, San Diego
Author for correspondence: Kerry McKenzie, kmckenzie@ucsd.edu

Abstract: A suite of questions concerning fundamentality lies at the heart of contemporary metaphysics. The relation of grounding, thought to connect the more to the less fundamental, sits at the heart of those debates in turn. Since most contemporary metaphysicians embrace the doctrine of physicalism and thus hold that reality is fundamentally physical, a natural question is how physics can inform the current debates over fundamentality and grounding. This Element introduces the reader to the concept of grounding and some of the key issues that animate contemporary debates around it, such as the question of whether grounding is 'unified' or 'plural' and whether there exists a fundamental level of reality. It moves on to show how resources from physics can help point the way towards their answers – thus furthering the case for a naturalistic approach to even the most fundamental of questions in metaphysics.

Keywords: fundamentality, grounding, physics, metaphysics, naturalism

ISBNs: 9781108714020 (PB), 9781108657617 (OC)
ISSNs: 2517-7273 (online), 2517-7265 (print)

Contents

1 Introduction

Fundamentality is the chief preoccupation of contemporary metaphysics. So much is clear from the fact that there has, in recent years, been an 'explosion' of new work aimed at articulating the concept of fundamentality and the cognate notions of ontological priority, ontological dependence, and grounding (Berker 2018: 732; Kovacs 2019; Schaffer 2016: 50). But it is evident also in the explicit identification of metaphysics, by contemporary metaphysicians, with the study of the fundamental in particular. Consider, for example, the following quotes.

- Sider (2011: 5): 'The heart of metaphysics is the question: what is the world ultimately, or fundamentally, like?...The truly central question of metaphysics is that of what is *most* fundamental...'
- Lowe (1998: 2): 'Metaphysics has traditionally been thought of as the systematic study of the most fundamental structure of reality – and indeed, this is the view of it that I would like to support'.
- Paul (2012a: 4): 'Metaphysics concerns the search for general and fundamental truths about the world...[it] wants to determine the natures of the world, especially the fundamental natures'.
- Schaffer (2009: 379): 'Metaphysics as I understand it is about what grounds what. It is about the structure of the world. It is about what is fundamental, and what derives from it'.
- Bigaj and Wüthrich (2016: 8): 'In a nutshell, metaphysics is the study of the fundamental structure of reality'.

While sometimes presented as the 'traditional' view of the subject (Dorr 2010), this is by no means the only way that metaphysics has been encapsulated historically.[1] Another characterization one frequently encountered, for example, was as 'the most general attempt to make sense of things' (Moore 2011) – a characterization that sits awkwardly with the idea it is principally concerned with a privileged subset. Nevertheless, it is evidently as *the study of the fundamental* that forms the conception with which many contemporary practitioners self-identify. Together with other commitments, however, it is a conception that generates an internal tension. For *physicalism* is arguably a highly orthodox stance within philosophy generally – indeed, has been described as being 'as close to a bit of orthodoxy as one will find in contemporary philosophy' (Hall 2010). Correspondingly, physicalist metaphysics today

---

[1] See Loux (2006: 2–10) for an instructive discussion of different characterizations of metaphysics.

constitutes 'the dominant metaphysics of nature' (Seager 2016: 7). Since physicalism is usually glossed as the idea that 'fundamentally everything is physical' (Jaworski 2011: 71), one might wonder how it is that physics and contemporary metaphysics could possibly be different, and yet also fit together. For how can it be that it is metaphysics that is definitionally concerned with the fundamental, while according to metaphysicians themselves it is physics that is assigned the task of describing fundamental domains? And how can it be that physics makes such scant appearance in current work in the metaphysics of fundamentality, if what metaphysicians are ostensibly most interested in theorizing about is the subject matter of that field?

A standard response to this question of disciplinary dovetailing is that metaphysics is charged with articulating the concept or intension of 'fundamentality' and its cognates; physics, on the other hand, is charged with uncovering its extension. Thus while, given physicalism, it may be the job of physics to tell us what the fundamental entities are, it is the job of metaphysics to tell us *what it is* to be fundamental in the first place. This answer is a natural continuation of another traditional view in which metaphysics is centrally concerned with the question of what the ontological categories are and how each of them ought to be characterized (Loux 2006: xi); the empirical sciences, by contrast, are tasked with identifying the members of these categories (most saliently for science, those of object, property, and law). Since proponents of this picture argue that nothing can be so much as individuated absent clarity on what category it belongs to, the work of the metaphysician is regarded as *conceptually prior* to that of the empirical scientist (Lowe 1998; Paul 2012a). And what goes for categories presumably goes for fundamentality. For how can physicists identify the *fundamental* members of those categories without a prior grasp of what it is to be fundamental, and so without a definition supplied by metaphysicians?[2]

Thus, on a broad but recognizable view of metaphysics in a physicalistic setting, physics tells us what the fundamental laws, objects, and properties are, while metaphysics tells us *what it is* to be a law, object, or property, and *what it is* to be fundamental. Such a picture of course goes a long way to explain why there has been such scarce engagement with physics in the metaphysics of fundamentality. As a general view about the relations between

---

[2] L.A. Paul is perhaps the most explicit recent exponent of the view just outlined: see Paul (2012a, 2012b). I should say that Paul also states that input from science is not irrelevant to metaphysics: she says, for example, that 'the division of labor is not sharp' (2012b: 12) and that 'science can force a revision of category theory on metaphysicians' (2012b: 15). But I do not see how this observation is compatible with (what seems to be) her core claim that 'the metaphysician should be concerned to *prescriptively develop and understand* the prior, deep, and general truths about the fundamental natures of the world' (2012a: 6, italics added), and so presumably impose a one-way normative constraint on science.

these two fields, however, it is one that I find deeply problematic – certainly insofar as it is taken to sanction the identification of one field as methodologically or conceptually prior to the other. My reasons stem simply from the very mainstream view that the method of 'reflective equilibrium' is essential to philosophy in general and to metaphysics in particular (Lewis 1983: x–xi). According to this method, philosophical claims of a general nature should be regarded as continually subject to revision and adjusted to fit our confident beliefs about particular states of affairs. And since one source of beliefs about particular states of affairs is empirical data, this means that physics must be capable, in principle at least, of having some kind of substantive input into the highly general statements concerning how things are that are routinely made by metaphysicians. Of course – in naturalistic metaphysics at least – we have already seen this dialectic play out with respect to some of the familiar metaphysical categories. Quantum mechanics, in particular, arguably spells deep changes with respect to what we mean by 'object', 'property', and 'law' – changes that may have a domino effect through our metaphysics as a whole.[3] But if the progress of physics can cause ruptures in our concepts of the ontological categories, why not with our concepts of fundamentality? Given how much of what has seemed indubitable, necessary, or obvious to metaphysicians regarding fundamental ontology has been upturned by physics, should we not be open to the idea that physics can upend some a priori claims made about the nature of fundamentality itself, and so be deserving of a 'seat at the table'?

For these reasons, the ways in which physics might inform contemporary metaphysical debates surrounding the concept of fundamentality has long been a focus of my work.[4] As is so often the case in metaphysics, these debates take the form of dualities. Is the fundamental a concept that should be taken to be primitive, or can it be analyzed?[5] If the latter, can the fundamental be characterized in intrinsic terms, or must it make reference to the special relation it bears

---

[3] For Schrodinger's views on how quantum mechanics forces us to give up classical conceptions of objects, see Bitbol (2007). For a nice discussion of how quantum mechanics changes our concept of law, without dispensing with it completely, see Born (1956). For an argument that gauge theories radically upset the way we can think about properties, see Maudlin (2007: chapter 3). For a lengthy statement of how changes such as these can reverberate across metaphysics as a whole, see French (2014).

[4] For example, McKenzie (2017a) argues that constraints in high-energy physics undermine that the fundamental may be thought of as 'brute and inexplicable'. McKenzie (2017b) discusses how dualities in physics can question the idea that fundamentality is an essential feature.

[5] Wilson is a proponent of the former view (see Wilson 2014: 560); others claim her argument rests on an equivocation (Raven 2017), or neglects the distinction between a property and its bearer (Mehta 2017).

vis-à-vis the non-fundamental?[6] If this latter, relational conception is right, can the fundamental be defined in wholly positive terms, or must it be defined in terms of its *lack* of grounding, dependence, or whatever relation is analytically central?[7] And so on. But any such project must first acknowledge that are all sorts of good reasons to understand the fundamental, at least at the outset, in terms of its relation to an assumed superstructure of non-fundamentalia (and this even if we decide that the ladder may ultimately be kicked away).[8] These reasons are both metaphysical and epistemological. They are epistemological, in that it is almost universally agreed in physics that we do not yet have a grasp on what is in fact fundamental, so that our current ideas concerning the fundamental are formed through a kind of extrapolation from facts pertaining to what we assume to be non-fundamentalia. And they are metaphysical, in that unless we want to prejudge the answer to the question of whether there must of necessity exist a 'fundamental level' then the more neutral starting point in a metaphysics of fundamentality is surely with the relative concept. For these reasons, the right place to start with any investigation of fundamentality is with a specification of 'ontological priority', and what we mean by the 'hierarchy of levels' taken to structure of the world – a notion that, as we will see, is multiply ambiguous. As such, it is a naturalized metaphysics of 'levels' and 'priority' – that is, of relative fundamentality – that is the focus of the discussion to follow.

This is therefore somewhat inevitably a work about 'grounding' and the further naturalization of grounding metaphysics. Grounding has after all emerged in the last decade or so as the generic name for the 'level-connector' in metaphysics and so as a notion intimately connected to the notion of ontological priority.[9] Now I will concede at the outset – as this vaguely apologetic tone no doubt conveys – that some will regard this as an unnecessary and even misguided focus for a work that claims to be in the business of further naturalizing the metaphysics of fundamentality. The notion of grounding has after all been rather mocked by some philosophers of science – philosophers who are otherwise happy, in some ways at least, to talk of both levels and priority – as a pretentious and borderline vacuous addition to the current stock of concepts in

---

[6] Sider (2011) holds the former; proponents of grounding-based definitions of fundamentality the latter.

[7] On how these two relate, but might also come apart, see Leuenberger (2020).

[8] This thought is expressed in Schaffer (2010: 36): 'Anyone who is interested in what is fundamental–where to be fundamental is to be ultimately prior–must understand some notion of priority'.

[9] While Fine (2001) is regarded as a landmark work in the field, both Sider (2020: 748) and Berker (2018: 733) refer to the focus on grounding as a phenomenon of 'the past decade'.

metaphysics.[10] There are also (as we will see) self-identified metaphysicians who are deeply sceptical about recent moves to reify an antecedently unrecognized relation of 'grounding' – or as those sceptics often put it, 'big-G' grounding – understood as a highly abstract and generic relation subsuming more specific relations of priority and theorizable largely independently of them. But while some of what I will argue here has implications that favour the sceptics in that debate, it seems to me that the emergence of a metaphysics of grounding, and the move towards 'hyperintensional' frameworks in metaphysics more generally (Nolan 2014), has been such a constructive development within the metaphysics of science that it seems appropriate to frame my discussion of levels and priority in those terms.

This Element begins with a brief overview of some of the literature on grounding, and moves on to consider how attention to physics can illuminate some key questions surrounding (i) the nature and (ii) the structure of the grounding relation. Thus, Section 2 outlines some of the broad points of agreement in metaphysics regarding how grounding is to be understood, before giving a snapshot of some of the contemporary debates around it. While there are many, the main focus will be that over whether there is some unified relation of 'grounding' that may be regarded as the unambiguous referent of grounding talk. Section 3 will then argue that this question should be answered in the negative, for considerations of how composition is understood in physics suggest that we must recognize two notions of 'metaphysical explanation', and hence two notions of grounding itself. This lends support to the view, also argued for there, that we should understand 'the' hierarchy of levels as a two-dimensional structure – something that in turn suggests it is wrongheaded to regard metaphysics as somehow 'more fundamental' than physics (as some contemporary practitioners are apt to suggest). In Section 4, I consider the question of whether the relation of grounding should be regarded as necessarily well founded – that is, of whether we should embrace the doctrine of 'foundationalism'. Drawing on recent work on vicious regresses and a case study from the history of particle physics, I draw a cautious conclusion that infinite regresses of grounds – and even vicious such regresses – should be sanctioned in a naturalistic metaphysics, and so foundationalism rejected. Section 5 is a brief conclusion.

Before I begin, however, let me make two clarifications with regard to my aim of naturalizing the metaphysics of grounding. The first is that I am not alleging that physics has thus far had no input into that literature, as to do so would be to exaggerate. Most obviously, the phenomenon of quantum

---

[10] This is based more on experience than on anything that I can find in the literature. But see here: https://scientiasalon.wordpress.com/2015/03/02/metaphysics-and-lack-of-grounding/.

entanglement makes frequent entry into that debate, and this is largely through the work of Jonathan Schaffer (whose presentations of the subject will loom large in the discussion later in the text). But it nevertheless seems fair to me to say, with Bryant, that the grounding project is for the most part 'a rather non-naturalistic enterprise — a messy hotbed of unchecked metaphysical speculation and a fine example of...metaphysics only nominally constrained by science' (Bryant 2018).[11] Thus, while this is by no means the only such attempt, there is arguably plenty of room in the metaphysics of grounding for further input from physics. The hope is that this Element represents some of that further work.

The second point I want to underline is that while I hope this work makes even clearer than before the value of looking to physics to illuminate fundamentality metaphysics, no claim is made to the effect that non-naturalistic or a priori metaphysics is devoid of value. I mention this because the last fifteen years or so have borne witness to much dispute about the relative value of (what has been called) 'armchair', 'analytic', or 'a priori' metaphysics versus more 'scientific' or 'naturalistic' approaches, with James Ladyman and Don Ross going so far as to state outright that the former 'contributes nothing to human knowledge' and as such ought to be 'discontinued' (Ladyman and Ross 2007: vii).[12] Nothing so incendiary will be alleged here. On the contrary, since this work takes analytic metaphysics as its starting point and then uses it to illuminate something interesting *about physics*, if the naturalistic metaphysics that results from this process has value then presumably so does the former.[13] The moral is simply that when it comes to the metaphysics of fundamentality, as with so much else, it is dialogue that points a way forward.

2 Getting on the Dartboard with 'Levels', 'Priority', and 'Grounding'

As already noted, there has been an explosion of interest in what may be called 'stratified metaphysics' in recent years. By this is meant metaphysical projects centred on the pre-theoretic intuition that the world is in some sense 'hierarchically organized'. In this picture, vast portions of the world are 'derived from',

---

[11] Note that Bryant herself says this is 'not entirely the case' – which is of course consistent with saying that it largely is.

[12] Note that I am not going to agonize in this work over the definition of these ambiguous and somewhat contentious terms, as little in my discussion really hangs on the distinction. But see Guay and Pradeu (2020) and Chakravartty (2017) for illuminating discussions of the ambiguities that swirl around the distinctions between 'naturalistic' and other forms of metaphysics.

[13] The ethos of this work is thus that of the 'Toolbox' approach of French and McKenzie (2012, 2016).

'produced by', or 'brought about by' the remainder, and in that sense need not be taken as metaphysically 'brute'. The remnant, by contrast, has a privileged status in the world, being in some way 'responsible' for all else that is in it, and for that reason deserves a privileged role in our theorizing. To commit to this general *gestalt* is to commit to the idea that there exists, seemingly as a matter of objective metaphysical fact, some kind of *order of priority* in the world, with the less fundamental portions existing *in virtue of* the remainder.[14] However it is that this edifice is to be conceptualized, clearly some kind of ordering relation is required to make sense of it. And at the heart of the contemporary understanding of this relation of ontological priority is the notion of 'ground'.[15]

While the idea of priority in nature is of course an ancient one, lying behind the recent explosion of work on grounding is an enhanced self-consciousness of at least two facts. The first is that the modal notions frequently reached to by the previous generation of metaphysicians to express ideas about priority are ill-equipped to do so adequately. As such, there has been a widespread feeling that something 'new' is necessary, opening the door to a novel and hopefully fertile research programme centred on something different. Sider sums up the motivations for the move nicely.

> Metaphysics has always needed a 'level-connector'. One doesn't get far in metaphysics without some sort of distinction between fundamental and non-fundamental facts, or between more and less fundamental facts... We've flirted with various ways to connect the levels: meaning, *a priori* entailment, supervenience. But consider the connection between the high-level fact that New York City is a city and the underlying physical reality–some fact that involves the global quantum state, suppose. This connection is clearly not a matter of meaning in any ordinary sense; language per se knows nothing of quantum mechanics. Nor is it *a priori*. Supervenience is a step in the right direction since it's a metaphysical (rather than epistemic or semantic) account of the connection between levels, but it too is inadequate...

> So there's a niche for a metaphysical but nonmodal conception of the connection between levels. That niche has been filled by ground (Sider 2020: 747–748).

The principal reason cited for why modal notions are inadequate to capture concepts of priority is that they are too 'course-grained' to capture explanatory

---

14 Not every attribution of fundamentality in philosophy connotes something worldly. To take just one example, from Popper (1962: 322): 'A theory like logic may be called "fundamental," thereby indicating that, since it is the theory of all sorts of inferences, it is used all the time by all sciences'.

15 As is standard, I use 'ontologically prior to' and 'more fundamental than' as interchangeable.

notions.[16] By contrast, the 'in virtue of' locutions that pepper pre-theoretic discourse about where things belong in the edifice of priority are taken to be constitutively explanatory in character. And it is the enhanced self-awareness that philosophical disagreements are often over 'in virtue of'-type claims that is the second reason for why grounding has taken on such a large presence in the field. Given that such ostensibly diverse debates as dualism versus materialism in philosophy of mind, positivism versus naturalism in philosophy of law, consequentialism versus deontology in moral theory, and conventionalism versus realism in aesthetics, may each naturally be construed as a debate over what properties occur in virtue of what, it isn't difficult to argue that *entire schools of thought* in philosophy rest on ideas about priority.[17] Underlying much of the recent flourishing of literature is the sense that investigation into 'in virtue of' locutions has yet to be undertaken in a way that does justice to their omnipresence across the discipline. Here, for example, is Audi:

> The phrase 'in virtue of' is ubiquitous in philosophy. Nearly as pervasive are the protestations that it is poorly understood and in need of clarification. Far less common are sustained attempts to elucidate this phrase and its philosophical significance. I propose that it expresses a primitive, noncausal relation of determination, which I call grounding (Audi 2012: 685–686).

As the quote indicates, the explanatory relation taken to be at work here is assumed to be non-causal.[18] A stated reason for this is that while 'causation links the world across time, grounding links the world across levels' (Schaffer 2012: 122). In explicit contrast to causal explanations, then, these 'level-connecting' explanations are taken to denote synchronically co-existing relata and are by now dubbed 'metaphysical explanations'. Indeed, grounding is often presented as *the* relation of 'metaphysical explanation', as though to give a metaphysical explanation *just is* to give a statement of grounds. Consider, for example, the following quotes.

- 'I take grounding to be metaphysical *explanation*: to say that ϕ grounds ψ is to say that ϕ provides a metaphysical explanation of ψ ' (Litland 2013: 20).
- 'Some truths metaphysically explain, or ground, another truth just in case the laws of metaphysics determine the latter truth on the basis of the former' (Wilsch 2015: 1).

---

[16] As Koslicki (2015: 308) puts it: 'any such purely modal relation is too explanatorily coarse-grained to capture and illuminate the nature of the connections at issue'. On explanation being a hyperintensional notion, see Nolan (2014).

[17] For a sense of the range of applications, see e.g. the list of questions in Raven (2015: 323).

[18] This is almost uniformly agreed on, although Alastair Wilson defects (Wilson 2018a).

- 'A number of philosophers have recently become receptive to the view that, in addition to scientific or causal explanation, there may be a distinctive kind of metaphysical explanation, in which explanans and explanandum are connected, not through some sort of causal mechanism, but through some constitutive form of determination or "ontological ground"' (Fine 2012: 37).

These, then, give some of the reasons behind the 'explosion' of literature on grounding: modal notions are too coarse-grained to carve out the hierarchy of levels and relations of priority, and the notion of 'metaphysical explanation' that we must reach to instead is underdeveloped. And this notion of 'metaphysical explanation' operating across levels is intimately tied up with the notion of ground – perhaps even being identical with it. As one might expect from philosophy, however, the proliferation of work on grounding owes also to widespread disagreement, along virtually every dimension, on how grounding is to be understood. Nevertheless, several core theses stand out as relatively fixed points in the debate, and as such are frequently presented as 'the orthodoxy' (Bliss and Priest 2018; Maurin 2019). We can use these relatively fixed points to get a better grip on the notion: more detail will be added as needed later.

1. WORLDLY AND OBJECTIVE.[19] Grounding links up entities in the world, and those links are themselves parts of the world independent of how we think about it. As such, grounding obtains between worldly entities, and the obtaining of grounding relations is an objective fact.[20] Thus, whether molecular properties ground solidity, whether laws ground their instances, and whether physical states ground mental states are facts about the world that long predated us, and that hold independently of our concepts, interests, and theorizing. As Maurin puts it, 'according to the "orthodoxy" grounding is a hierarchical dependence-relation that holds between worldly facts or states

---

[19] That grounding obtains as a matter of objective fact is rarely stated outright as an orthodox assumption (although see Audi 2012: 691). However, it is at least implicit in much – indeed the vast majority – of the extant grounding literature. Thus while he himself challenges the objectivity assumption, Dasgupta (2017: 89) writes: 'The issue here concerns whether constitutive explanations [i.e. explanations in terms of ground] are objective, or whether they are relative to facts about us such as our interests or concerns; facts that may vary from culture to culture or time to time. As I use the term, a "realist" thinks the former while an "anti-realist" thinks the latter.... Recent discussions of ground tend to tacitly presuppose a realist picture–indeed much of the literature largely overlooks the possibility of anti-realism'.

[20] One debate I am not going to get into here is over whether grounding relations can take on entities of any category (as Schaffer [2009] argues) or rather solely facts about those entities (as in Rosen [2010]). I will write as though grounding can obtain between objects, properties and laws, as well as the facts about them; it seems to me that anyone who disputes this assumption is free to translate such talk into talk of facts alone.

of affairs. …[I]t is an objective and mind-independently obtaining [relation].. always holding between what is less and what is more fundamental' (Maurin 2019: 1574).

2. ENTAILMENT.[21] Grounding is a relation of determination, such that the existence of the grounds entails, as a matter of metaphysical necessity, the existence of the grounded. This is often expressed as the 'entailment principle' (Rosen 2010: 118). Grounds are therefore *metaphysically sufficient conditions* of whatever is grounded in them, and establishing that the grounds of some phenomena are instantiated is therefore enough to infer that the grounded phenomenon is instantiated as well.

3. LOGIC. The logic of ground is a non-monotonic, strict partial order, always directed from what is less to what is more fundamental (Maurin 2019: 1574; Rabin 2018: 38). Thus, grounding is asymmetric, irreflexive, and transitive, and to say that x grounds y is to imply that x is *more fundamental* than y.

4. EXPLANATION. Grounding bears an intimate connection to metaphysical explanation. For to say that x grounds y is to commit to some kind of claim to the effect that y occurs *in virtue of x*, and thus that x *explains y*; further, as we have seen, the mode of explanation involved in connecting levels is standardly categorized as *metaphysical explanation*. What is meant by the notion of 'metaphysical explanation' is now the principal locus of debate in this area. But '[m]inimally, most seem to agree, it is an explanation which accounts for the nature and/or existence of something with reference to something else on which the first thing *non-causally* and *synchronically* depends. Something which somehow (again, non-causally and synchronically) *determines* or *makes* the second thing exist and be the way it is' (Maurin 2019: 1574).

5. FOUNDATIONALISM.[22] Relations of ground necessarily bottom out into a set of ungrounded fundamentalia. As this is often put, grounding relations are necessarily 'well-founded'.

Taken together, these principles define a picture which 'looms large over contemporary analytic metaphysics: a picture according to which reality is hierarchically arranged with chains of entities ordered by relations of ground

---

[21] Skiles (2015) notes that this is 'orthodoxy', although he himself contests it; Bliss and Trogdon (2019: section 5) call it the 'default' view. Likewise Cameron (2019) holds that it is 'widely held'.

[22] Bliss and Priest (2018) and Rabin (2018) both include this as a standard assumption. Note, however, that there are some otherwise 'orthodox' theorists that explicitly dispute this (e.g., Gideon Rosen). But see Section 4 for further textual justification that this is the orthodox view.

and/or ontological dependence terminating in something fundamental' (Bliss and Priest 2018: 1).

The rationale for the final assumption here will be the explicit target of Section 4. For now, however, let me say a little about the other components of the orthodoxy, beginning with assumption (2), ENTAILMENT. The motivation for this is the idea that relations of grounding are relations of determination, which seems perfectly intuitive: for the very idea behind stratified metaphysics is that once some things are settled, enough has been done to (in some sense) 'bring about' the rest. The more fundamental in that sense *determines* all that remains about the world. (As Audi puts it, '[l]ike causation, the relation will be one of determination, roughly, of making so, of bringing about, or of being responsible for' [Audi 2012: 688]). That it is primarily determination we have in mind when we think about priority is also clear from the 'all God had to do' gloss that is frequently put on the fundamental (Wilson 2018b: 499) – a cliché which at least makes clear that the fundamental is pretheoretically understood as a sort of *sufficient* condition on everything else that there is.[23]

For all that seems obvious now, however, it actually took us a while to be clear on the idea that the more fundamental is to be understood in these terms. Indeed, until recently, it was more common to find priority expressed in terms of ontological *dependence*, and so the non-fundamental as that which in some way depends for its existence on something else.[24] It was also common to find 'determination' and 'ontological dependence' used as interchangeable terms.[25] This is a problem, however, as it seems clear that these are different concepts. Most obviously, thinking of the prior as that which determines the posterior would seem to make the former a sufficient condition of the latter; thinking of the prior as that on which the posterior depends, however, would seem to make it a necessary one. But it goes without saying that necessary and sufficient conditions are not in general the same thing. Moreover, one may argue that the two notions are not co-extensive, as there are examples of entities that are determined by others but not ontologically dependent on them; there are also examples of entities that are dependent on others and yet not determined by them.[26] But most dramatically of all in the current context is the fact that

---

[23] Schaffer's 'blueprint for reality' taps into the same idea (Schaffer 2010: 39).

[24] While no longer endorsed as the right one, it seems this conceptualization has a long and distinguished pedigree. It is, for example, that which seems to be operative in both Aristotle and his teacher: 'Some things then are called prior and posterior... in respect of nature and substance, i.e. those which can be without other things, while the others cannot be without them, – a distinction which Plato used' (Aristotle 1984: 1688).

[25] See McKenzie (2020a: section 3) for some explicit examples.

[26] Cases of multiple realizability provide instances of the former, and entanglement relations in quantum mechanics are a stock example of the latter (McKenzie 2020a). See also Audi (2012: 692–693); Barnes (2018).

dependence and determination have directly contradictory implications for how entailment connects to priority. Since to say that x ontologically depends on y is (at least in part) to say that the existence of x entails, as a matter of metaphysical necessity, that y exists as well, x's necessitation of y would make x the more fundamental of the two on a determination-based conception, but the less fundamental on the dependency-based one.[27]

Determination and dependence, then, are not the same relation – nor is one even the converse of the other. And we can say this while granting that there are close connections between the two notions – most obviously, that anything a non-fundamental entity ontologically depends upon will necessarily occur in its grounds, and hence be part of what determines it. As I have argued elsewhere, given that it is determination that gets to the heart of our pre-theoretic conception of the fundamental, it seems that it is this that should win out as the way that we understand ontological priority.[28] Among other innovations, this clarification makes space for a view in which the fundamental can be ontologically dependent, as in 'coherentist' metaphysical views.[29] Given that the relevant distinctions here were elided for so long, the clarifications that have been gained through recent discussions represent a compelling example of progress in metaphysics.[30] But the point for present purposes is that ENTAILMENT gets things the right way round.

Turning now to assumption (3), LOGIC, we should note first that as stated it is divided into two parts: a part stating that grounding is an ordering relation and a part stating that it is directed from the more to the less fundamental. Both of these assumptions seem appropriately packaged together: it is after all its role in generating a hierarchy of being that largely motivates the introduction of grounding, and it is hard to make sense of the notion of a hierarchy without some ordering being essential to it. But the first assumption has been subject to critique: there is in fact by now a cottage industry in metaphysics directed at providing counterexamples to the logical principles supposed to govern ground. Indeed, by now, 'every component of the orthodoxy has been challenged' regarding the logic of ground (Rabin 2018: 38). For example,

---

[27] This point is also made in Calosi 2020: 15–16.

[28] Note that even with that clarification, there remain debates over whether the fundamental should be regarded as that which determines everything else or as that which is itself undetermined. While the former is often presented as the 'completeness' characterization and the latter as the 'independence' characterization, in either case it is determination not dependence that is being used to define fundamentality. (See Leuenberger [2020] for a discussion of the connections between these two notions.)

[29] There are a number of arguments for the feasibility of coherentist metaphysics, and indeed for its actuality. See e.g. Calosi and Morganti (2018) for an example of the latter.

[30] Progress here is marked by the fact that, for all the historical conflations, '[a]t this point it is fairly standard to distinguish between grounding and ontological dependence' (Kovacs 2019).

Schaffer (2012) questions transitivity; Jenkins (2011) and Barnes (2018) are routinely presented as questioning irreflexivity and asymmetry, respectively. Here I should note that many of these supposed counterexamples to the logic of grounding in fact consist of relations of *ontological dependence*, not relations of determination, failing to exhibit partial orderings. (The paper by Barnes frequently cited in this context is called 'Symmetric Dependence', and Jenkins' is 'Is Metaphysical Dependence Symmetric?'.)[31] Thus, at the very least, more work would need to be done to show that these results translate into the grounding-as-determination context. Schaffer's argument against the transitivity of grounding has also been roundly criticized (see e.g. Litland 2013).

Rather than wade into these debates here, however, I want to flag only that Gabriel Oak Rabin makes a compelling case that *even if* such counterexamples succeed, they do not seriously challenge the idea that grounding outlines the contours of a hierarchical ordering (Rabin 2018). He argues, for example, that much of the threat to an ordering posed by instances of symmetry or reflexivity can be deflected by switching from a 'Simple Principle', according to which *x is at a lower level than y if x grounds y*, to the 'Slightly Less Simple Principle', according to which *x is at a lower level than y if x grounds y and furthermore y does not ground x*. Where the second condition fails, entities should in most cases simply be placed on the same level – something that does not endanger the idea that there are levels at all. As such, I will not say much about these supposed counterexamples to the logic of ground here. Even if they succeed, it seems they do not do violence to the idea that grounding can be conscripted in the articulation of stratified metaphysics.

Focus now on the second aspect of what I have called LOGIC, and so on the directedness of the grounding relation. It is this that allows us to move from an ascription of grounding to one about relative fundamentality. This connection between grounding and priority, together with some rhetoric in the literature, can leave one with the impression from that if one has a theory of ground then one automatically has a theory of ontological priority sufficient to build up the 'hierarchy of being'. But on the face of it, this is not true. To see this, note that all LOGIC asserts is that a claim that *x grounds y* is a sufficient condition for saying that *x is more fundamental than y*. But it is much less clear that grounding links between two entities are necessary for us to regard one as more fundamental than the other. (E.g., one might instinctively regard an atom of lead as more fundamental than the contents of a can of nursery paint,

---

[31] Berker (2018: 757) states that most of Barnes' examples can be reinterpreted as symmetries of ground. But I am not so sure.

even if – one hopes – lead atoms do not count among the paint's grounds.) It therefore seems that we are, pre-theoretically at least, happy to place entities in relations of priority even in the absence of grounding relations between them. This is because it seems that – to cite Rabin again – 'a complete story of the world's grounding relations…gives us both more and less than we want from the layered conception' (Rabin 2018: 40). It gives us less because it doesn't suggest any obvious way of attributing priority relations to pairs of entities, like the atom and the paint, unconnected by relations of ground. And it gives us more because much of our priority talk is at the level of general types, whereas grounding relations are typically understood to relate particular instances (see e.g. Audi [2012: 692]). As such, LOGIC does not settle how to build an analysis of priority out of grounding.

The first extended attempt in the recent literature to build up an analysis of priority from a notion like grounding may be found in Karen Bennett's *Making Things Up* (Bennett 2017: chapter 6). This approach is based on a comparison of the 'number of steps' required to 'build' entities up from the fundamental level and aims at allowing relative fundamentality attributions between different kinds of entities unrelated by connections of ground. However, while valiant, even she admits that the details here are going to be difficult to flesh out – not least in the circumstance in which there is no fundamental level to start from. Since then, new proposals partially inspired by Bennett's work have appeared in the literature (see e.g. Correia [forthcoming]; Werner [2020]). However, I myself am not going to discuss these proposals here. This is in part because the arguments I want to offer in the next two sections, and that I hope shed new light on grounding, are independent of whether a full analysis of priority in terms of grounding is possible. But it is also because some of the desiderata on such an analysis seem to be as much in question as the analysis itself.[32] For these reasons, while in what follows I will focus on grounding in an effort to get a grip on aspects of the hierarchy of priority, it should be understood that only some aspects of the world's priority structure will be illuminated in so doing.

Turning now to assumption (4), EXPLANATION, it is arguably this feature of ground that has contributed most to the explosion of interest in it. And its close connection with explanation is perhaps also the most essential feature of ground: indeed, as one can find remarked on in the literature, it is probably the only aspect of grounding that is not disputed (Glazier 2020: 121; Maurin 2019: 1574). This feature has obvious connections to feature (2), ENTAILMENT: for

---

[32] For example, is a theory to be preferred if it makes the hierarchy of priority a total order? Or should it be the case that some entities are simply incomparable with respect to how fundamental they are? Some at least are happy with the latter idea, as we will see later.

while explanation is a hyperintensional notion, it is nevertheless – and thinking here most obviously of the deductive-nomological model – intimately connected with the notion of entailment. However, EXPLANATION in fact sits rather uncomfortably with (1), OBJECTIVITY, and (3), LOGIC – making the grounding 'orthodoxy' a rather unstable one (Thompson 2016).

To see why, we should note first that as stated EXPLANATION is something of a banality: the devil will be in the details, and there are at least two ways that the 'intimate connection' between grounding and metaphysical explanation may be understood. On one – sometimes dubbed the 'unionist' view (Raven 2015) – grounding is simply to be *identified* with metaphysical explanation. Fine, Dasgupta and Rosen all count as unionists, as conveyed in remarks such as these:

- 'A number of philosophers have recently become receptive to the idea that, in addition to scientific or causal explanation, there may be a distinctive kind of metaphysical explanation, in which explanans and explanandum are connected…through some constitutive form of determination. I myself have long been sympathetic to this idea of constitutive determination or "ontological ground"' (Fine 2012: 37).
- 'As I use the term, "ground" is an explanatory notion: to say that X grounds Y just is to say that X explains Y, in a particular sense of "explains" (Dasgupta 2014: 3).
- 'The grounding relation is an explanatory relation–to specify the grounds for [p] is to say why [p] obtains' (Rosen 2010: 117.)

Other by contrast hold the 'separatist' view, according to which relations of ground are what justify or 'back' explanations (or conversely, thought of as that which metaphysical explanation 'tracks'). According to this view, '[e]xplanation is not and does not account for grounding – on the contrary, grounding is what makes possible and "grounds" explanation' (Rodriguez-Pereyra [2005: 28]; see likewise Audi [2012: 687–688]).

One reason to favour the latter, 'backing' view is that we tend to think of explanations as things that cognitive agents may exchange with each other – something clearly not the case with, say, the relations of ground that exist between laws and the objects they govern. Note, however, that if we choose to adopt this 'separatist' route, then the motivation for LOGIC becomes less clear – for these formal properties are frequently ascribed to the relation of grounding precisely because they are properties of good explanations (Maurin 2019: 1576; Raven 2015: 689). For example, non-monotonicity reflects the fact that good explanations should avoid the inclusion of irrelevant information, and irreflexivity reflects the worthlessness of explanations that are circular. But if we go the 'unionist' route and identify grounding with explanation, then it is now the rationale for OBJECTIVE that becomes murky. For we often think of explanations in subject-involving terms, insofar as their success or aptness in a given context is a function of an agent's cognitive interests. Thus even if the explanations in question can be thought of as 'worldly', why think that they should qualify as objective?

This dilemma posed by the debate over whether grounding is identical with or rather backs explanation reveals how awkward the 'orthodoxy' really is. Nevertheless, it is not a debate I am going to get entangled with here, although for the sake of simplicity I will talk in what follows as though the 'identical with' side has things right. A debate I do want to wade into, however, is another dichotomous debate over the nature of metaphysical explanation: that over whether grounding ought to be regarded as 'unified' or 'plural'. For while not every debate in the literature around grounding is to my tastes, patience, or purposes, *this* is a debate that I regard as having tangible metametaphysical and naturalistic implications.

Unfortunately, as I suspect all parties in this debate will concede, what is even meant by grounding being 'unified' or otherwise is not easy to articulate precisely (Koslicki 2015; Raven 2017). In order to keep things moving, here we will focus on the strongest, most straightforward, and I suspect most popular interpretation of the matter.[33] According to this interpretation, grounding is unified if, when we make attributions of grounding in the various different philosophical contexts, we are talking of fundamentally one and the same relation (Berker 2018: section 5). As such, on this view, in whatever context we take there to be an instance of metaphysical explanation, there is present, perhaps in addition to other relations, one and the same relation of ground. Clearly, if this is the case, then we can hope that our talk of grounding is unambiguous across different contexts, and as such that it successfully refers in each. And if that is so, then it seems grounding can be discussed at the level of generality that characterizes the recent literature. But if, by contrast, there are in fact 'several distinct relations of grounding or dependence in the vicinity, and [...] uncritical invocation of "the" grounding idiom conflates them' (Rosen 2010: 114), then it is entirely unclear that grounding theorists are succeeding in doing what they think they are doing – namely, producing a theory that is at once illuminating, highly general, and that succeeds in picking out something in the world. As such, the debate over the 'unity of ground' amounts to a dispute over the very legitimacy of the 'grounding revolution'.

---

[33] A fuller range of options is presented in Bennett (2017: section 2.3); Koslicki (2015); Raven (2017).

To see why there is scope for worry here, recall that a major motivation for work on the notion of grounding is that 'in virtue of' talk crops up all over philosophy. One can, for example, find grounding talk utilized in debates over the status of configuration spaces in quantum physics, as well as in those concerning the relation between aesthetic objects and the institutions that display them; one hears of determinables grounded in determinates, disjunctions being grounded in their disjuncts, the mental as grounded in the physical, and the normative as grounded in the natural.[34] As such, it seems that virtually any pair of entities of any category can stand in relations of ground.[35] Moreover, while the same term 'grounding' may be deployed in each case, the 'specific metaphysical relations' that motivate or flesh out such talk can differ across these contexts – with some kind of logical determination relation being appropriate in the disjunction case, and maybe functional realization in the mind–body context. These more 'specific metaphysical relations' that motivate attributions of grounding Jessica Wilson calls the 'small-g' grounding relations (Wilson 2014). These latter are contrasted with what she calls 'big-G grounding' – the relation which is supposed to be 'ultimately at issue in contexts in which some goings-on are said to hold "in virtue of," be (constitutively) "metaphysically dependent on," or be "nothing over and above" some others' (Wilson 2014: 535), and thus operative in *any* context in which a small-g relation is attributed. Talk of 'big-G' grounding thus presupposes that these different small-g relations all have something significant in common – namely, at least on the strongest interpretation, that whenever we talk about any of them we are somehow *also* talking about it. But now consider the sheer variety of contexts in which 'in-virtue-of' locutions are used. Given this variety, how much content can we realistically expect to be associated with whatever it is that they have in common? There is surely a worry that there can be little significant binding the various small-g relations together – perhaps nothing but a loose formal analogy. But if *this* is the case, then there is a worry that the relation of 'big G' grounding that supposedly makes a cameo in all these varied contexts is actually just as coarse-grained as the purely modal notions whose inadequacies motivated its introduction. Among other things, this suggests we are not dealing with an entity 'intimately connected' with explanation after all.

For these reasons, the supposed 'grounding revolution' gives rise to the worry that 'by treating a collection of phenomena which is in fact

---

[34] For an example of the quantum case, see North (2013); for the aesthetic case, see Abell (2012).

[35] This way of putting it assumes that entities of any type can stand in relations of grounding, not just the facts about them. But the moral will transfer to that less liberal notion. (E.g. Koslicki (2015) attacks Rosen, who is a 'facter', just as much as Schaffer, who takes entities of any category to stand in relations of ground.)

heterogeneous as though it were homogeneous, we have, if anything, taken a dialectical step backwards' (Koslicki 2015: 307). A further worry is that approaching metaphysical explanation in such an abstract way results in purely self-generated problems – such as the vexing question, animating much debate in the literature, of whether grounding itself is grounded (see references in Wilson [2018b]). As Wilson puts it:

> That the question 'What Grounds Grounding?' is a spandrel question generated by Grounding's overly abstract 'nature' is supported by the fact that no comparable question arises when the operative understanding of physical dependence is instantiated with one or other small-g relation…. [By contrast], if someone tells you that mental states are determinables of physical determinates, there's no temptation to ask, 'But in virtue of what do they stand in the determinable–determinate relation?' (Wilson 2018b: 508).

This debate, to me, is a particularly interesting one because in a way it strikes at the heart of the question of what it is that we want out of metaphysics, and thus at questions of its value. And it is of course questions of value that have been at the heart of recent skirmishes, briefly alluded to in Section 1, between naturalistic and a priori metaphysicians. For metaphysics, at least on some renderings, aspires to be 'the most general attempt to make sense of things', and in that sense aims towards more abstract descriptions of the nature of the world than are offered by other disciplines. But the cost of generality can of course be banality, and so it seems that what metaphysics aims to do is to somehow 'walk the line' between these two extremes – carving out a unique and uniquely insightful take on the world in the process. The core claim made by Wilson and Koslicki seems to be that the grounding project has tipped too far towards the former end of the spectrum to count as meaningful intellectual activity, and may in fact – as critics of metaphysics in general have long alleged – be simply generating problems for itself for which there is no way out. What we are seeing here, then, is a kind of 'insider' scepticism about metaphysics that nevertheless channels familiar positivist themes.[36] Insofar as naturalistic metaphysicians, such as myself, like to think that our metaphysics has superior epistemic credentials compared to that which have been repeatedly attacked historically, it seems to me that *this* is a debate in which we have some investment.

Given the importance of this debate to contemporary metaphysics there has of course been pushback against the objections raised by Wilson and Koslicki. Defenders of 'big-G' grounding may point out that the very notion of

---

[36] This language of 'insider' and 'outsider' sceptics about grounding metaphysics is taken from Hofweber (2009).

'small-g grounding' *presupposes* that there is some reason to lump all the various instances together, thus betraying the existence of a common feature or features that entitles them to be regarded as relations of priority. All the 'big-G' grounding project is doing, they will say, is attempting to articulate what that is. Furthermore, the mere fact that the small-g grounding relations form a diverse list with respect to the relata they take does not itself mean that there is only a loose formal analogy between them, or – more strongly – that the relations between them differ *qua metaphysical explanations*. As Audi puts it,

> Even the view that there is only a generic similarity, that there is a different species of noncausal determination at work in each case, strikes me as under-motivated. What differentiates the species? If it is only that one concerns normative properties, another determinables, still another dispositions, this does not yet give us a reason to think that *how the determination* works differs in each case, simply because it relates different kinds of fact. So I take the burden of proof to be on those who think there are different relations at work to show why, to show in what way the determination differs in the different cases' (Audi 2012: 689).

Koslicki objects, however, that the 'differentiations between the species' that she has in mind transcend mere differences between the nature of the relata they take on, and that we do have reason to think that 'the way the determination works' differs between instances of small-g grounding relations. For example, the standard assumption that disjunctions are grounded in their disjuncts shows that overdetermination of grounds is permitted in some cases; in the case of determinate-determinable relation, however, such overdetermination is thought to be metaphysically impossible. This seems to be a difference in terms of *how* things ground as well as in what it is that is grounded. Similarly, and as Audi himself admits, the grounding of the wrongness of an act in its being an act of lying may be subject to 'extenuating circumstances', while any appeal to such circumstances seems inapplicable in paradigm non-moral grounding cases. Concerning as it does *when* grounding relations between moral and natural facts obtain, Koslicki notes that this looks like 'a difference in "how the determination works", and not just a difference in the nature of the relata' (Koslicki 2015: 324). Complicating this response in turn, however, is the fact that some deny that the relations of semantic determination involved in the disjunction case should be regarded as instances of grounding at all – undermining the relevance of the former example to a discussion of the logic of metaphysical structure (see e.g. McSweeney [2020]). And Koslicki also concedes that 'the details of the moral/natural case' are 'very complex issues [not] settled by

[her] very brief remarks' (Koslicki 2015: 324). As such, it is not clear how far this debate as it stands has developed beyond a trade over intuitions of where the 'burden of proof' should be taken to lie.[37]

For all its importance, then, if there is one thing regarding the debate over whether grounding is 'unified' or 'plural' is that it is in a bit of a tangle. Nevertheless, I think we can say something definitive here and without wading too deeply into the sorts of questions that currently characterize the debate. For I think there are resources in elementary physics that show that the sceptics here are right – or at least, that they are right if we take seriously the Humean analysis of laws (to be explained later). The reason is that we must recognize a plurality of relations of metaphysical explanation if we are to circumvent a version of the worry regarding explanatory circularity known to arise in the context of Humeanism. Since these relations are identified with the relation of ground, we must recognize a plurality of the latter also.[38] To see how this argument for pluralism works, we must first be clear that claims about a 'hierarchy of levels' occur in two different contexts within contemporary metaphysical discourse: one that is focused on the relations of priority that occur between the members of categories, and that are typically studied in science; and one that is focused on the relations of priority that occur between the categories themselves, and that are typically studied in metaphysics. I will argue that these two hierarchies are not the same or even related as part to whole; rather, each hierarchy defines a *different dimension of priority*. This is a point with potentially wide-ranging metametaphysical implications, and so to the argument we now turn.

3 Levels of Nature, Levels of Metaphysics, and the Plurality of Priority

We know that the relation of grounding is regarded as the 'level-connector' in metaphysics. And part of what motivated its introduction was the acknowledgement that within philosophical discourse talk of 'levels' is ubiquituous. As we will see, however, it seems there are at least two distinct notions of 'levels' that are discernible in the metaphysics literature. The first, which I will call 'levels of nature', are the levels that are identified by science; as we will see, they are typically linked by relations of priority holding between the members

---

[37] In addition to the aforementioned Audi quote, see Rosen (2010: 114); Schaffer (2010: 689).

[38] While what I've just said assumes the 'identical to' view of the relation of ground and explanation, it seems safe to assume that the same conclusion would follow on the 'backing' view. For if the relevant metaphysical explanations differ then presumably whatever backs them must differ too.

of a given category. The second, which I will call 'levels of metaphysics', are the levels that are identified in metaphysics; in paradigm cases, these are linked by relations of priority holding between the categories themselves. By bringing together reflections on how composite objects are explained in physics with the Humean analysis of laws of nature, I will argue that there are reasons to see these two relations as distinct, and moreover distinct *qua* relations of explanation. This has implications for the 'big-G grounding' debate, and hence also for how the very project of theorizing grounding ought to be conducted. It also has big-picture implications for the relation between physics and metaphysics. I'll begin by discussing 'levels of nature'.

3.1 Levels of Nature

While by no means the only or most sophisticated example, a useful touchstone for our purposes regarding talk of 'levels of nature' is Jonathan Schaffer's *Is there a fundamental level?*, written in 2003 (and so a few years before the 'grounding revolution' really got going).[39] As the title suggests, this paper focuses on the question of whether the hierarchy of levels of nature should be regarded as 'bottoming out' into a fundamental one. As he put the matter then:

> Talk about 'the fundamental level of reality' pervades contemporary metaphysics. The fundamentalist starts with (a) a hierarchical picture of nature as stratified into levels, adds (b) an assumption that there is a bottom level which is fundamental, and winds up, often enough, with (c) an ontological attitude according to which the entities of the fundamental level are primarily real, while any remaining contingent entities are at best derivative, if real at all... (Schaffer 2003: 498).

Thus the 'levels' talk here is explicitly intended as an example of what we can call 'stratified metaphysics'; following Schaffer himself, I will denote the levels that are in focus here 'levels of nature'. Schaffer continues by identifying what he takes to be the 'central' and 'peripheral' connotations of the levels metaphor in this context:

> The central connotation of the 'levels' metaphor is that of (a) a *mereological structure*, ordered by the part-whole relation. This connotation is

---

[39] It is a useful touchstone because it connects the case study of this section with that of the next, which focuses on anti-fundamentalism. But it is useful also because it seems to me that it is solely this paper's questioning of the existence of a fundamental level that is regarded as controversial. This gives us reason to regard his other assumptions regarding the levels hierarchy as largely reflective of the broader field.

implicit both in Newton's talk of 'the smallest particles', and in Cough-
lan and Dodd's use of the cliché 'fundamental building blocks'; and this
connotation is implicit both in Oppenheim and Putnam's use of part-whole
relations to distinguish microtheory from macrotheory (p. 407), and in Kim's
presentation of the fundamentalist picture…

The peripheral connotations of 'levels' include those of (b) a *supervenience
structure*, ordered by asymmetric dependencies; (c) a *realization structure*,
ordered by functional relations; and (d) a *nomological structure*, ordered by
one-way bridge principles between families of lawfully interrelated properties. Those who speak of levels typically suppose that most if not all of these
connotations comport. Thus William Lycan speaks of a 'multiple hierarchy
of levels of nature, each level marked by a nexus of nomic generalizations
and supervenient on all those levels below it on the continuum' (p. 38). Of
course these connotations can come apart, and if they do, or at least to a
significant enough degree, then perhaps the entire 'levels' metaphor is best
abandoned (Schaffer 2003: 500).

Thus, there are here 'mereological, supervenience, realization, and nomological connotations of "levels"' – that is, of 'levels of nature'.

Several points deserve mention here. First, this being 2003, supervenience
is still in the running as a relation of priority (a 'level-connector'). But by 2010
the same author was to claim, and in what seems to be currently the most cited
paper on grounding, that supervenience 'fakes' priority structure.[40] Thus, were
the same paper to be written today this instance of a 'level-connector' would
almost certainly not make the cut, and as such we will ignore it here. Second,
the mereological conception of the levels hierarchy is here presented as the
paradigm, just as it was in Oppenheimer and Putnam's classic discussion of
the hierarchy of sciences. While this assumption is widely shared among metaphysicians, it is an assumption that some philosophers of physics and physicists
themselves regard as outdated and naive. Brown and Ladyman, for example,
write that '[l]ike materialism, mereological structures are obsolete philosophical conceptions in the face of modern physics, and have lost credibility and
utility in the effort to describe reality' (2009: 28). While likely not the only
motivation, underlying this thought may be the idea that our concept of 'parthood' connotes particles, whereas the fundamental entities posited in current
physics are fields; particles are at best localized excitations of fields, and as
such derivative and non-fundamental.

I myself do not really feel the force of this objection, however, and as
such am going to continue to regard relations of parthood among objects to

---

[40] At time of writing, the paper has 1,361 citations.

stake out relations of priority.[41] The reason is simply that it strikes me that it continues to be *very much* the business of physics to account for composite entities in terms of simpler ones – such as protons in terms of quarks – whether the entities concerned are particles *or* fields. Certainly the discourse of quantum field theory is replete with talk of 'fundamental' and 'composite' entities, and in which moreover the two terms are used contrastively.[42] And even if talk of composite fields is ultimately elliptical for talk about the corresponding (and less fundamental) particles, since the particles are themselves part of the levels structure it seems to me that compositional structure represents an aspect of the world that any naturalistic metaphysics is obliged to capture.

The third point I want to emphasize about Schaffer's presentation just now is that questions of what in fact *stand* in these relations – that is, of what objects compose what, of what properties realize what, and of what laws may be derived from what – are taken to be questions that it is primarily the business of *empirical science* to investigate. This seems clearly correct: not even the most ambitious of enquirers would affect, after all, to have discovered the quark structure of protons from the armchair. And finally – and most importantly for our purposes – it seems that when we talk about 'levels of nature' what we are typically doing is staking out relations of priority among *members of the same category*.[43] That is, we are talking about what objects are prior to what objects; or what properties are prior to what properties; or what laws are prior to what laws. In that sense, we may say the levels of nature are *intra-categorial levels*. Now to be clear, it is *not* the case that what *makes* a member of a given category posterior to some other member or members of that category is necessarily a function of *members of that category only*. Deploying some standard terminology of the grounding literature, it may be that members of the category occur only as *partial* grounds within a *full* ground that includes entities of other categories (e.g. Fine [2012]); indeed, it will be argued below

---

[41] Or at least, I will where it is unambiguously the case that the composite objects concerned have their properties accounted for by the properties and relations of their parts. All my examples will be of this type. I suspect that the argument will work just as well if directed at the hierarchy of properties as opposed to that of objects.

[42] To take an example in the case of particles chosen more or less at random, from *Reviews of Physics* in 2019: 'Ever since Rutherford began to probe the structure of the proton, the question of whether or not the particles we observe are fundamental or composite is a perennial question. Investigations of quark compositeness are not fundamentally different than the Rutherford experiment, and involve investigations of the number of high-mass quark-quark interactions' (Rappocio 2019: 11). Fields are likewise routinely bisected into the categories of 'elementary' or 'composite' (see e.g. discussions of composite Higgs field models in 'beyond the Standard Model' physics).

[43] This is not just a feature of the examples given: the very fact that there is a worry about comportment between the various intra-categorial hierarchies implies that this is the case.

this is so in the case of (at least) the grounds of composite objects. But being a partial ground within a full ground of an entity x is universally taken as sufficient for being more fundamental than x, so this is sufficient for the member of the category concerned to be prior to the grounded entity of the same category (e.g. Rosen [2010: 116]).[44] It seems, then, that when we talk about 'levels of nature' we are foregrounding a particular aspect of the priority structure that obtains between the grounded and its grounds – namely, that which obtains between the grounded entity and others of the same category. While these relations may not exhaust the complete grounding story, they are no less real for that.

Let us therefore denote the priority relations defining levels of nature as P_n. A first stab at a picture expressing the 'levels of nature' would then look something like Figure 1, where in each case the entities:

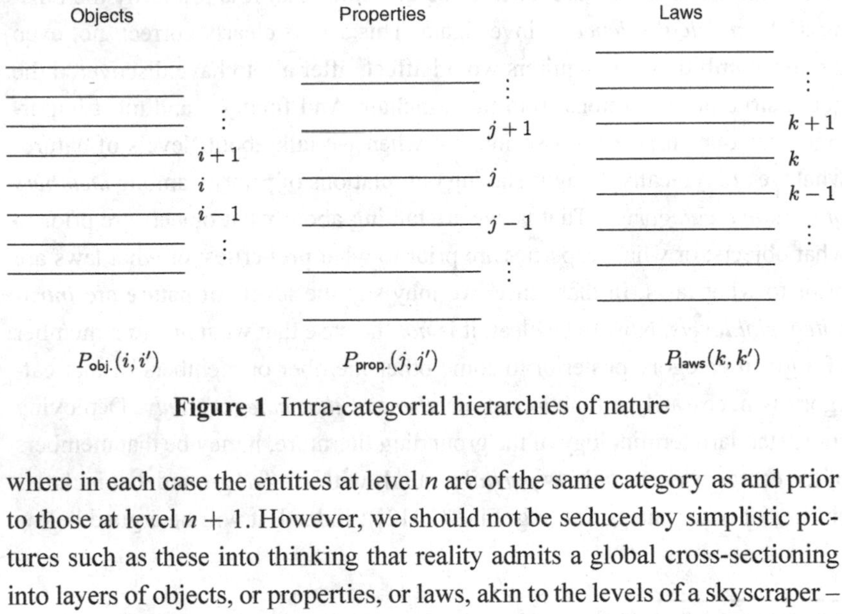

Figure 1 Intra-categorial hierarchies of nature

where in each case the entities at level n are of the same category as and prior to those at level $n + 1$. However, we should not be seduced by simplistic pictures such as these into thinking that reality admits a global cross-sectioning into layers of objects, or properties, or laws, akin to the levels of a skyscraper – that is, a cross-sectioning such that for any given entity of any given category there is an unambiguous answer to the question of which 'level of nature' it belongs to. For while composition (we are taking it) stratifies objects into (what deserve to be called) 'levels', it is not the case that we can compare any two objects with respect to relations of parthood. For example, a neutron gas is composed of free neutrons, and an atom is composed of electrons, protons, and neutrons bound together in different ways; while we may unambiguously say that both the atom and the gas are located at a level above the neutron,

---

[44] Or I should say, almost universally: see Fine (2001: 27).

given that neither the atom nor the gas is a part of the other it seems we as yet lack the resources to answer the question of how their respective levels relate.[45] Viewed in this way, talk of levels tracks 'piecemeal explanatory achievements' rather than 'monolithic' entities that 'reach across all of nature' (Craver 2014: 12). However – and though here I repeat myself – to my mind there is nothing that this more restricted application of the language of levels that makes the hierarchy of levels of nature any less objective or significant: one can deny that levels are monolithic in this sense and still be able to claim that someone who thinks that temperature, or solidity, or conductivity must be taken as primitive in our physical inventory alongside more fundamental properties has genuinely missed something about the world, and of which the scientific significance cannot be overstated.

What now of the question, raised at the close of the last quote from Schaffer, of how each of these levels 'comport' – that is, of the extent to which the levels of one intra-categorial hierarchy are approximately isomorphic to those of another? One thing that seems fairly clear right away is that we can at the very least expect there to be significant correlations and correspondences between the levels of each hierarchy: for example, part of our reason for regarding water to be constituted by H_2O molecules is that aggregates of the latter lawfully behave as water does (Loewer 2012: 131). But as Schaffer also rightly notes, we do not expect to find strict 'comportment' or isomorphism between each hierarchy. For example, both realizing and realized properties may belong to one and the same object, meaning that objects at one level can instantiate properties from different levels (cf. Melnyk 2003: 21). Further – to take an obvious example – if the true law of gravity (whatever it should turn out to be) is both universal and fundamental then it seems that fundamental laws can apply to non-fundamental objects just as well as to fundamental ones. However, demanding that there be neat comparisons between categories with respect to their 'levels' structure as a condition of speaking of levels at all is clearly too strong a requirement on the meaningfulness of levels talk. Perhaps most saliently, given that we do not even require of a single category that there be an unambiguous answer of whether or not two entities in it are on the same 'level', it is not even clear what 'strict comportment' between the different intra-categorial hierarchies could possibly mean. Therefore I myself see

---

[45] Assuming that to be *at a less fundamental level* just is to be *less fundamental*, the issue here is much the same as that discussed in Section 2 in connection with Bennett's attempt to analyze priority in terms of building (a relation much like grounding). Just as we did there, we can note that we might deny that the priority structure of the world is given by total order while still hoping to make at least *some* priority claims regarding entities not connected by relations of ground.

no clear reason to infer from lack of comportment that the metaphor 'should best be abandoned', as Schaffer suggests, as opposed to simply understood as a metaphor that engenders rich complexity. This is fortunate, for as Craver notes,

> The suggestion that we might be better off abandoning the levels metaphor is about as likely to win converts as the suggestion that we should abandon metaphors involving weight or spatial inclusion. These metaphors are too basic to how we organize the world to seriously recommend that they could or should be stricken from thought and expression (Craver 2014: 2).

For all those qualifications on comportment, however, it will nevertheless be convenient to superimpose the three pictures above into one hierarchy of 'levels of nature' as represented in Figure 2. What is intended by this amalgamation is simply that for each of the different categories posited in science – the categories of object, property, and law – their members may be systematically connected by relations of priority. In the remainder of this section I am going to represent things as if there is, as a matter of fact, a fundamental level of nature, and I will understand this to contain all of the fundamental objects, fundamental properties, and fundamental laws. The status of such an assumption is something that will be critically examined in the next section. But I assume it here because it paves a more efficient route to my conclusion: the general moral of this section will not hang on it.[46] It makes it easier since we may safely presume that such a level, if it exists, features *only* fundamental objects, fundamental properties, and fundamental laws, and so the worries about the lack of comportment between these intra-categorial hierarchies do not arise. Furthermore, the Humean analysis of laws typically assumes such a level, and it is that analysis that turns the handle of the argument of Section 3.3.

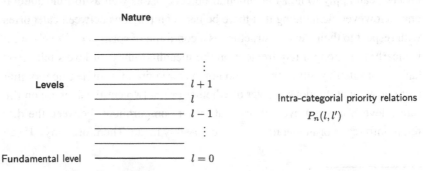

Figure 2 Schematic representation of levels of nature

---

[46] For example, nobody to my knowledge defends the idea that there must *not* be a fundamental level to reality. And if there are any worlds with both laws and a fundamental level, then the moral will stand.

This completes our brief discussion of 'levels of nature'. Such levels are carved out by relations of priority between the members of a given category, and in that sense are 'intra-categorial' constructions; they are discovered by empirical means. They do not 'reach across all of nature', but for all that they reflect an objectively existing priority structure, and so something real: a metaphysically and scientifically significant aspect of the world.

3.2 Levels of Metaphysics

There is more, however, to the 'levels' discussed in metaphysics than those we take to be identified by science. For as have noted, a stated motivation for the 'grounding revolution' was the growing realization that many *distinctively metaphysical* questions were at their heart questions about priority. Thus Schaffer, in his highly influential paper *On What Grounds What* published in 2009, claims that '[o]n the now dominant Quinean view, metaphysics is about what there is' (Schaffer 2009: 347). After arguing that such a paradigm was incapable of making sense of many (and for Schaffer, clearly most) of the 'central metaphysical questions' he urges the 'revival of a more traditional Aristotelian view, on which metaphysics is about what grounds what' (Schaffer 2009: 347). As such, this metaphysics 'involves concepts one will not find in Quine or Carnap' (Schaffer 2009: 354). Thus he writes (adding boldface to help things along momentarily):

> The philosopher raised on the Quine-Carnap debate who turns to the central metaphysical questions will leave confused. She will find debates such as: (i) metaphysical realism versus idealism, (ii) realism about numbers versus constructivism, (iii) realism about **universals versus nominalism**, (iv) **substratum versus bundle theories of objects**, (v) dualistic versus materialistic theories of mind, (vi) **substantival versus relational theories of space**, and (vii) monistic versus pluralistic theories of the cosmos. She will find little disagreement about what exists, but profound dispute over what is fundamental (Schaffer 2009: 362).

All of these, says Schaffer, are debates over priority, that cannot so much as be expressed in the 'now dominant Quinean' framework; hence some kind of 'neo-Aristotelian' paradigm is required. On this he writes:

> …[T]he neo-Aristotelian will conceive of the task of metaphysics as:

> *Aristotelian task*: The task of metaphysics is to say what grounds what.

> That is, the neo-Aristotelian will begin from a hierarchical view of reality ordered by priority in nature. The primary entities form the sparse structure

of being, while the grounding relations generate an abundant superstructure of posterior entities. The primary is (as it were) all God would need to create. The posterior is grounded in, dependent on, and derivative from it. The task of metaphysics is to limn this structure (Schaffer 2009: 351).

As MacBride notes, this 'unfavourable comparison of Quine to Aristotle rests upon caricatures of both' (MacBride and Janssen-Lauret 2015: 298). And others have argued that the 'dominance' of Quinean metaphysics is greatly overstated – something that undermines the supposedly revolutionary character of what is going on (Wilson 2014: 539). But the present issue is not the novelty of this conception of metaphysics but rather the centrality of priority to it. And what I want to emphasize here is that, at a minimum, questions (iii), (iv), and (vi) are all centrally concerned with *relations of priority between categories*. The issue of realism versus nominalism about universals, (iii), is standardly framed in terms of whether it is the category of objects or that of properties that qualifies as fundamental.[47] Of (iv), he writes:

> Moving to the debate over substrata as per (iv), both the substratum and bundle theorists accept the existence of objects and properties. The dispute is over priority. For the substratum theorist, objects are prior, and properties are dependent modes…For the bundle theorist, properties (be they universals or tropes) are prior…Objects are then bundled out of compresent property complexes (Schaffer 2009: 362–3).

Although he himself does not put it in quite this way, the debate in (vi) over the substantival nature of space concerns the priority of objects (whether 'material' or space-time points) over their topological and metric relations – thus accounting for its centrality to the debates over ontic structural realism (see e.g. Ladyman and Ross [2007: Chapter 3]).

We see, then, that many 'central metaphysical questions' concern priority, and specifically the priority relations between categories – most saliently in these cases, the categories of object and property/relation. However, to Schaffer's list, we must surely add another – namely, the debate over the fundamentality of laws.[48] This, after all, is one of few debates in metaphysics that even philosophers of science otherwise sceptical about metaphysics are often willing to engage in. And here the situation is much as Schaffer describes it, with few in this debate outright denying the existence (or even the objectivity) of laws. Rather, what is at stake is whether laws (as per the Humean view) may be regarded as deriving from facts about the behaviour of objects, or whether

---

[47] Certainly this is the case with resemblance nominalism: see e.g. Hakkarainen (2015: 87).

[48] Indeed, some take Schaffer (2008) to have initiated the integration of the literatures on grounding and Humean laws: see e.g. Kovacs (2021).

(as per e.g. Maudlin's 'primitivist' view) they are to be taken as unanalyzable and fundamental (cf. Loewer [2012, Section 2]). Thus while the priority relations definitive of 'levels of nature' paradigmatically concern the members of a given category, it seems that a very large part of what metaphysics is tasked with is arranging into the categories themselves into an order of priority. Since it is the business of metaphysics, on a view like Schaffer's, to take the entities postulated in science and prise them into their various categories, let the levels staked out by more and less fundamental categories be denoted 'levels of meta-physics'. Correspondingly, let the priority relations connecting these levels be denoted P_M.

What we call 'stratified metaphysics', then, contains reference to two ostensibly different hierarchies: the hierarchy of nature and the hierarchy of metaphysics. To complete the metaphysical picture, then, we need to think about how these two hierarchies relate. How, if at all, do the levels of one intersect with those of the other? How do P_M and P_N, which define those levels, connect with one another? How, in particular, are the directions defined by P_M and P_N 'aligned' with respect to one another (the notion of directedness being essential to a hierarchy after all)? Given the different relata of each hierarchy, these questions do not have obvious answers.

To help us answer them, return to Figure 2 – the hierarchy of levels of nature under the assumption there is a fundamental level. There we explicitly assumed that *in this level* we find the fundamental objects, fundamental laws, and fundamental properties.[49] And recall that a critically important question – indeed, for metaphysicians perhaps the most important question – is that of what category or categories qualify as most fundamental. Now clearly, the most natural place at which to direct this question is the *fundamental level of nature*: certainly we will have to ask it here if we have to ask it anywhere. Thus, we must ask, of the fundamental objects, fundamental laws, and fundamental properties, which belongs to a fundamental category? To make things concrete, suppose for argument's sake that the category of laws comes out as non-fundamental and that of objects as fundamental, as e.g. many Humeans will hold.[50] Then it will be correct to claim that even the fundamental law of nature (the law of

---

[49] We will revisit this assumption later.

[50] Humeans could choose to analyze properties in terms of objects or vice versa. David Lewis – the chief architect of modern Humeanism – is himself a class nominalist and thus takes concrete objects as fundamental. As Loewer (2007: 314), notes: 'Lewis doesn't say much about the possible fundamental concrete entities that can occupy a space-time. Putative examples are the points and regions of space themselves, classical point particles, Bohmian point particles, one-dimensional strings, two-dimensional branes, electromagnetic fields, gunk, and wave function'. But given the generic nature of the argument to follow regarding bound states I don't think ambiguity on this point matters much.

quantum chromodynamics [QCD], say) is *non-fundamental*, and *less funda-mental* than the objects whose behaviour accords with it (namely quarks and gluons). We find ourselves in a position, then, in which the fundamental laws come out *as fundamental as* the fundamental objects (insofar as both populate the fundamental level of nature), but also as *less fundamental* than the funda-mental objects (insofar as they are members of a less fundamental category). On pain of contradiction, then, it seems we must index these priority claims to a *different* relation of priority in each case. And given that we could draw the same conclusion if another pair of categories instantiated at the fundamental level stood in that relation of priority with respect to P_M, it seems we cannot *analyze* P_M in terms of facts about P_N.

What this seems to confirm, then, is that we must recognize two priority relations, neither of which can be regarded as reducible to the other, in our total picture of stratified metaphysics. This circumstance we can call the 'plurality of priority'. Given the seeming analytical independence of the two notions, as far as I can see the only way to represent the different priority claims we want to make is by carving out *another dimension* of priority along which the fundamentality of the entities of different categories found at the fundamental level of nature can differ. The corresponding picture (or at least part of it), then, presumably looks something like this:[51]

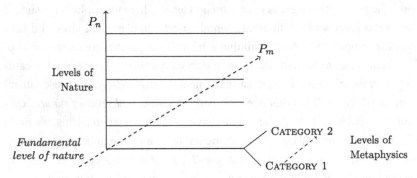

Figure 3 The two-dimensional hierarchy of priority

where here I understand P_m, and hence the two levels of categories, as directed 'into the page'. The two priority relations we recognize in metaphysics, then, ought to be regarded as carving out different dimensions of priority. And since direction is essential to any notion of hierarchy, it seems to me that we should

---

[51] There are other ways to argue for this same conclusion. For example, some take it to be defin-itive of metaphysical categories that they find application at each level of nature (see e.g. Imaguire [2020]). If so, the overall picture would have a scheme of categories 'attached' to each level of nature as they are here with the fundamental level. But I postpone a fuller and more positive discussion of how the two sets of levels relate to another occasion.

regard the two hierarchies afoot in metaphysics as essentially distinct. What we talk about in stratified metaphysics, then, is not best described as 'the hierarchy of being', so much as a conjunction of (at least) two distinct hierarchies each defined with respect to a different relation of priority.

If this is indeed the right way to think about priority in metaphysics then the consequences are significant. First and most obviously, the multi-dimensional aspect of the stratification central to 'stratified metaphysics' is a structurally fundamental feature of it, and hence one of intrinsic metaphysical interest. But it is one that has not to my knowledge been emphasized. While we find the literature peppered with countless statements to the effect that reality possess 'an overarching layered structure' and that 'grounding is the relation by which the world is hierarchically structured' (Bliss and Trogdon [2019: Section 6.2]), there is little indication that 'the hierarchy' is thought of as having (what we might call) both 'verticality' and 'depth'. Indeed, talk of 'the axis of fundamentality' (Wilsch 2015: 3294) and of 'the ladder of metaphysical explanation' (Baron and Norton, 2019) suggests it is positively *not* conceptualized in this way. But it is of significant metametaphysical interest as well because it may be consequential for both how metaphysics is done and how its subject matter relates to that of physics.

To see this, we can note first that the plurality of priority may have consequences for how we draw inferences in fundamentality metaphysics. For 'stratified metaphysics' in general is often explicitly motivated by patterns observed in the sciences in particular. Our opening quote from Schaffer makes this apparent, as there it is stated that it is something about *levels of nature* that 'pervades contemporary metaphysics'. Bryant summarizes Schaffer's approach as follows:

> In the picture of the world that emerges from science, the world is structured by these dependencies [discovered by science], and this is the metaphysician's proper motivation for positing a hierarchical worldview structured by grounding relations (Bryant 2018: 2.2.1.).

But we can see something similar expressed right across the metaphysics literature. Here, for example, is Wilsch:

> Common sense recognizes a wide range of entities including tables and chairs, mental processes, social and moral institutions, and works of art, to only name a few. Science suggests that there are further entities, including atoms and molecules, sets and numbers, or organisms and species. Fundamental reality, on the other hand, contains only a few kinds of objects which correspond to the terms of the ideal fundamental scientific theory. So, if common sense and science provide accurate views of the world, the following question naturally arises: How does some austere fundamental reality give

rise to a dazzling variety of entities? How does the fundamental connect to the derivative?

It has recently been suggested that there is a univocal notion of metaphysical explanation, or ground, which connects the fundamental to the derivative [here citing, among others, Fine [2001, 2012]; Rosen [2010]; Schaffer [2009]] (Wilsch 2015: 3293).

But if there are a plurality of structures of levels involved in the overall picture then the extent to which conclusions drawn in the context of levels of nature have implications for priority in general is very much less clear. (We will return to this point at the end of the section.)

A further consequence, however, is that we cannot make sense of some recent rhetoric surrounding metaphysics and its relation to the sciences. For we noted in Section 1, and in connection with some worries about physicalism, that many metaphysicians hold that the work involved in defining the categories and the relation of priority is 'conceptually prior' to the work done by physicists. And it seems that at least some of these take this to go hand in hand with the claim that that which metaphysics describes is *ontologically* prior to that which science describes. L.A. Paul, for example, holds not only that 'many concepts of metaphysics are conceptually prior to the concepts of science', but also that metaphysics 'describes features of the world that are more fundamental than those of natural science' (Paul 2012a: 6). This is on account of the fact that '[t]he categories are metaphysically prior to the members of the categories' (Paul 2012a: 5). As such, she writes, 'physics tells us which fundamental (i.e. perfectly natural) physical objects, structures and properties are the actual members of the ontological categories, and metaphysics *takes it a level deeper*, by telling us what the fundamental categories are' (Paul 2012b: 234; italics added). And, as Alyssa Ney has recently pressed, such rhetoric is not at all unusual (Ney 2020).[52] Here for example is E. J. Lowe:

> The time has now come for me to offer my own answer to the question of whether, and if so how, metaphysics is possible. My view is that it is indeed possible: that is, I hold that it is possible to achieve reasonable answers to questions concerning the fundamental structure of reality – questions more fundamental than any that can be competently addressed by empirical science (Lowe 1998: 8–9; quoted in Ney 2020: 2).

---

[52] In this paper, Ney provides a battery of arguments as to why this claim cannot be right, at least not on any plausible ontological reading of 'fundamental'. I am in agreement with virtually everything Ney writes there; the argument to follow simply provides further support to her conclusion.

But insofar as science and metaphysics are each tasked with exploring different dimensions of priority then I do not see how we can make sense of claims of this sort. For presumably metaphysics could be said to be going to 'a level deeper' than science only if what it explores is located on the same hierarchy of levels. If, however, the relations of priority they invoke are 'orthogonal' to each other, as argued above, there is no sense to be made of this claim. Insofar as both are interested in determining some aspect of priority structure of the world, we can certainly see them as engaged in analogous and complimentary projects. But neither is obviously investigating something more fundamental than the other.

If the plurality of priority is a real feature of the world, then, it will be significant for both metaphysics and metametaphysics. Correspondingly, we should expect there to be pushback against the argument just given. One might object, for example, that – as presented at least – the argument rests on the idea that 'the fundamental level' contains entities drawn from several categories, and moreover categories that have been presumed to differ with respect to their fundamentality status. But one might question that in light of the conclusion just drawn. Perhaps, one might hold, what metaphysics does through its analysis of categories is effect a 'splitting' of the most fundamental level discoverable by science into a plurality of levels of nature, with (say) the fundamental level *of the hierarchy of objects* shown to be at a more fundamental level than that of the laws.[53] And if that is the case, then we arguably *can* make literal sense, of some sort at least, of the claim that metaphysics 'takes things a level deeper' than physics. Certainly, the argument for the plurality of priority just given could not get off the ground.

Now I myself do not think that this is a natural way to think about how the two sorts of hierarchies relate, or about what effect the analysis of categories has on the assumed contents of the fundamental level of nature. Humeans, for example, are perfectly happy to talk about 'fundamental laws' despite taking laws to be a non-fundamental category: indeed a frequent criticism of the Humean account is that it is *exclusively* 'an account of laws at the fundamental level', and so poorly equipped to handle laws at less fundamental levels (see e.g. Williams [2019: 225]). Furthermore, the idea that all of the fundamental properties, objects, and laws co-exist at the fundamental level receives expression, in one form or another, across the literature in more or less explicit forms and without taking sides on contentious questions regarding which categories are

---

[53] Here I can't help but mention that were this to be the case, there would be a nice formal analogy with fine structure in atomic physics, in which the property of spin splits what was assumed to be the lowest energy level of an atom into states of higher and lower energy. Alas, I don't think it is the case.

fundamental.[54] But I do not want to argue further for that here. Rather, what I want to do is support the plurality of priority via the claim that we should recognize a plurality of relations of metaphysical explanation. Since grounding and metaphysical explanation are so closely related, this argument also has direct implications for the supposed 'unity' of grounding and hence the 'big-G' grounding debate.

The argument for the plurality of relations of metaphysical explanation will be predicated on the Humean analysis of laws of nature together with the way that compositional structure – the 'central connotation' of the order of priority among objects – is generically explained in physics. I will argue that this plurality is necessary if Humeans are to circumvent a charge of explanatory circularity. The generic charge that the explanations Humeans offer are circular is not a new one, having been made already by (among others) Armstrong in the context of the Humean's explanations of the causal patterns found among events (Armstrong 1983: 40). The by-now standard response in that context, which was initiated by Barry Loewer, was to distinguish 'scientific' from 'metaphysical' explanation and break the circularity that way. What I will argue now is that this move cannot work in this context unless we are willing to distinguish two distinct kinds of the latter: one which connects the levels of nature, and the other the levels of metaphysics. The extent to which one must be a Humean to buy into this conclusion will be reflected upon at the end.[55]

3.3 Humean Laws and the Charge of Explanatory Circularity

Let's begin by recounting the ethos of the contemporary Humean account of laws of nature, chiefly associated with David Lewis.[56] Humeans in this tradition begin their analysis of laws with the assumed 'Humean mosaic', the manifold of space-time points, stretching everywhere and from the beginning to the end of time, endowed with the distribution of fundamental physical properties. This comprises the 'totality of facts' that it is the purpose of science to make sense of. Of course, the way that scientists make systematic sense of this vast totality

---

[54] Although they themselves argue against it, Hicks and Schaffer (2017) call this view the 'orthodoxy'. As an example they give North (2013: 186), who writes: 'How is a world built up, according to its fundamental physics? At the fundamental level, there is the fundamental ontology of the theory, there is the space in which this ontology lives, and there is some structure to that space. Then there are the dynamical laws, which say how the ontology evolves through this space over time'.

[55] I should note that Emery (2019) also argues that laws ground their instances and relates this to the circularity problem facing Humeanism. My own version of the argument focuses more specifically on instances of physical composition. Although doing so makes it vulnerable to certain objections to which hers is not, I think it makes more vivid case for laws grounding, as opposed to causing, their instances.

[56] For all his influence Lewis never wrote a dedicated paper on laws of nature. My own favourite exposition of the view for the uninitiated is Beebee (2000).

of facts is by subsuming them under laws of nature, and for the Humean it is this fact about scientific practice that points the way towards the analysis of what laws are. According to the Humean, the laws are the *axioms and theorems of the systematization of the totality of facts that strikes the best balance of simplicity and strength*. This, they will claim, respects Humean strictures and solves the problems plaguing other regularity-based accounts, all the while hewing closely to scientific practice.

There are of course numerous objections that can be made to this account of laws – not least what it means to be in the 'best balance' and whether this is a notion that can be accorded objective significance. But the objection that is relevant just now is that this analysis – for all that it claims to be inspired by scientific practice – in fact sits very uncomfortably with some core *explanatory* practices of science. While the objection can be found in an earlier form in Armstrong, Maudlin puts the point nicely.

> If one is a Humean, then the Humean Mosaic itself appears to admit of no
> further explanation. Since it is the ontological bedrock in terms of which all
> other existent things are to be explicated, none of these further things can
> really account for the structure of the Mosaic itself. This complaint has been
> long voiced, commonly as an objection to any Humean account of laws. If the
> laws are nothing but generic features of the Humean Mosaic, then there is a
> sense in which one cannot appeal to those very laws to explain the particular
> features of the Mosaic itself: the laws are what they are in virtue of the Mosaic
> rather than vice versa (Maudlin 2007: 172).

The point relayed here constitutes an objection because it is clear that scientists *do* use the laws to explain features of the mosaic *all the time*. Most saliently and typically, they use the laws plus facts about temporally earlier events of type e_1 to explain why it is that later events of type e_2 obtain. But since the law L has been determined in part by the fact that events of type e_1 are followed by those of type e_2, it seems the explanation of e_2 provided in terms of e_1 and L must be circular.

While acknowledging it as a serious objection if successful, Loewer defends the 'best system' view from this circularity objection as follows:

> I claim that this objection rests on failing to distinguish metaphysical expla-
> nation from scientific explanation. On Lewis' account the Humean mosaic
> metaphysically determines the L-laws [i.e. the laws according to Lewis].
> It metaphysically explains (or is part of the explanation together with the
> characterization of a Best Theory) why specific propositions are laws. This
> metaphysical explanation doesn't preclude L-laws playing the usual role of
> laws in scientific explanation (Loewer 2012: 131).

Thus the claim is that by distinguishing the type of explanation that obtains *between* the laws and the mosaic from the explanation the laws facilitate *about* the mosaic and the patterns of temporal succession in it then the circularity objection can itself be circumvented. The first form of explanation he designates as 'metaphysical' and the second as 'scientific'. Then the idea is that '[g]iven the distinction between metaphysical and scientific explanation the argument that L-laws cannot be involved in scientific explanations of one part of the mosaic by another falls apart' (Loewer 2012).

The heart of Loewer's defence against this 'long-voiced' objection to Humeanism, then, rests on a distinction between scientific and metaphysical explanation. What is also clear is that the paradigm of the former is taken to be causal explanation: Loewer writes, for example, that 'one of the signal differences between scientific and metaphysical explanation is that scientific explanations of particular events are temporally directed; prior events and laws explain subsequent events and not the reverse' (Loewer 2012: 131–2). By contrast – and as was emphasized in the last section – the latter explanations are 'synchronic' in character and explain the less fundamental in terms of the more. Now, we should note that it is generally accepted that causal and metaphysical explanation, understood in this sense, are fundamentally different forms of explanation. This is for good reason, as there seem clear disanalogies between the two types of explanation that are relevant to the metaphysics of priority. For example, it seems clear that we are not in danger of 'double-counting' if we say there is a cause in addition to its effect: there is no sense to the idea that the fire 'just is' the striking of the match together with the existence of the bundle of papers (not least because the fire might consume them and still be going strong). By contrast, we worry that we *are* doing so when we count the matchstick in addition to the 'particles arranged stickwise'; and since intuitions about 'nothing over and above'-ness are often recruited to express ideas about priority this earmarks a relevant difference between the two. Thus, insofar as scientific explanation equals causal explanation, the explanations scientists offer of patterns in the mosaic are different from those using such patterns to explain the laws, and the alleged circularity dissolves.

So far so good, then, for Loewer's defence of Humeanism. Complicating matters, however, is the fact that (as he concedes) Loewer does not possess a full analysis of this distinction, and in particular seems not to be committed to the idea that scientific explanation is restricted to causal explanation. He says only that '[s]cientific explanation of a particular event or fact *need not* show that it is grounded in a more fundamental event or fact but rather, *typically*, shows why the event occurred in terms of prior events and laws' (Loewer 2012: 131, italics added). This of course leaves it open that some scientific explanations may also be metaphysical explanations.[57] And whatever form of explanation it is that Loewer has in mind here in addition to causal explanation, given that we can assume scientific explanations can in general be expected to invoke laws there is a worry that *these* explanations, at once scientific and metaphysical, may not be immune to the circularity challenge.

What I want to argue now is that Loewer's implication here is correct: there *are* other features of the mosaic beyond relations of temporal succession that are explained by science, using laws, and which cannot be regarded as causal. Moreover, we have every reason to regard those explanations as 'metaphysical' in character, insofar as they explain the existence of non-fundamental entities in terms of the more fundamental. But *providing we recognize a distinction between forms of metaphysical explanation the Humean can rebut the circularity charge.* This, I claim, provides a reason to believe that the two forms of metaphysical explanation at work here are different *qua* explanations.

My argument is essentially a spatial analogue of that regarding temporally separated events that motivates Loewer to press the distinction between scientific and metaphysical explanation in the first place. The starting point for the argument is the fact that one thing we ask scientists to explain for us, in addition to why events of type e_1 are followed by events of type e_2, is why it is that certain *composite objects* are found to exist in the world. It is scientists, after all, who explain why the hydrogen atom – a composite of proton and electron – exists; why the helium atom – a composite of proton and neutron – exists; and why it is that there exist objects composed of a quark together with an anti-quark. These questions are non-trivial, as witnessed by the fact that plenty of a priori conceivable composite objects are *not* found amongst the furniture of the world. For example, while helium nuclei abound, it is thought that there are no structures composed of just two protons or two neutrons; and while one finds quark–antiquark pairs routinely popping out of accelerators, one never finds composites of two quarks or two antiquarks being produced. But this absence is something that physicists can explain.[58]

Now since composite objects are taken as a paradigm of non-fundamental objects, and since 'level-connecting' explanations relating the more to the

---

[57] By now the idea that there is much more to explanation in science than merely causal explanation is well taken: see e.g. Reutlinger and Saatsi (2018).

[58] The isospin symmetry of nuclear physics, together with the spin-dependence of nuclear laws, determines that *nn* and *pp* states are more weakly coupled than *np* states, and insufficiently strongly coupled to bind (see e.g. Bertulani and Danielewicz [2019, Section 4.2]). The fact that there are no quark–quark or antiquark–antiquark pairs follows from the fact they cannot be colour-neutral – a requirement taken to follow in turn from the 'confinement' property of QCD.

less fundamental are taken to constitute 'metaphysical explanations', it seems Loewer is quite right to leave open the possibility that scientists can provide metaphysical explanations – for it seems that they routinely *do*. Again, so far so good. But when we think about how these explanations proceed, and in particular how central laws are to them, it becomes clear that there is a need for Humeans to distinguish different forms of explanation within the genus of 'metaphysical explanation'.

At the core of this argument is the fact that what we are asking physicists to explain when we ask why certain composites and not others exist is why it is that some, but only some, sets of objects form *bound states*.[59] As the name suggests, bound states are entities that have a certain spatiotemporal profile: they are entities that retain a certain spatial proximity to one another over time as opposed to separating to an arbitrary degree. The paradigm objects of daily life clearly have this feature, and it was presumably partly on account of this that we were motivated to introduce the category of 'object' as the stable subject of predication in the first place. The practice continues in contemporary physics, however, with (for example) the 'particle zoo' discovered in the 1960s accelerators being identified with those combinations of quarks that form bound states.[60] However, despite the wholly quotidian nature of this fact that the parts of certain systems retain spatial proximity over time, from a physical perspective such systems form a special class and one does not expect an arbitrary pair or set of objects to exhibit it. It is this fact that some, but only some, combinations of entities have this spatially stable structure that cries out for explanation. Moreover, it is clear that this explanation will *simultaneously* be an explanation of an aspect of the levels structure of the world (insofar as it explains less in terms of more fundamental objects), *and* an explanation of an aspect of the space-time manifold or 'Humean mosaic' (insofar as it explains certain facts about spatial proximity). We should not be surprised that there is some connection between facts about spatial location and facts about priority: the metaphysician wants to say, pointing at the table, that there is priority structure *there* after all. It is this double-natured aspect of composite objects in physics that I claim mandates, for the Humean, differentiation within 'metaphysical explanation'.

To see why, we need to think more about how it is that the existence of bound states is explained within physics. The first thing to be clear on is that

---

[59] See McKenzie and Muller (2017) for a defence of the idea that composite objects reified in science are those whose components form bound states – thus furnishing a solution to the 'special composition question'.

[60] See Veltman (2018: chapter 8) for a nice introduction for the uninitiated to the explanation of the zoo in terms of bound states of quarks.

finding this explanation can be a highly non-trivial task even in ostensibly the simplest of cases. Take, for example, the simplest chemical entity, namely the hydrogen atom – an (in some sense) spherically symmetric entity composed of a single proton and a single electron with a radius of the order of 10^{-8}cm. The existence of this entity was experimentally established around the turn of the twentieth century and by now it constitutes a stock example in metaphysics of a composite object. The explanation metaphysicians provide for the existence of this entity will of course be in terms of the existence of its constituents – hopefully annotated with the fact that those constituents form a bound state. Thus Sider writes, perfectly innocuously, 'The fact that there exists an atom of hydrogen is grounded in the fact that there exist an electron and a proton bonded to each other' (Sider 2013: 761).[61] But scientists do not take the existence of this bonding relation to be matter of primitive fact: it is *why* neutrons and protons (and not two neutrons or two protons) are ever bonded to each other in the first place that forms the core scientific question. As such, the answer to this question will form part of the explanation of the existence of the less fundamental object – and hence, it seems, will form a part of its 'metaphysical explanation'.

The non-triviality of this why-question may be conveyed by the fact that, as was almost immediately apparent, the account of the electrostatic interaction that was given by the laws of classical mechanics was incapable of recovering the existence of the hydrogen atom as a bound state. The situation was nicely put in a 1915 textbook as follows:

> There seems to be no difficulty about the supposition that at very small distances the law of force is different from the inverse square. On the contrary, there would be a very real difficulty in supposing that the [classical electrostatic force] law $1/r^2$ held down to zero values of r. For the force between two charges at zero distance would be infinite; we should have charges of opposite sign continually rushing together and, when once together, no force would be adequate to separate them…Thus the matter in the universe would tend to shrink into nothing or to diminish indefinitely in size. The observed permanence of matter precludes any such hypothesis…
> (Jeans 1915: 168).

Such a consequence was physically disastrous. Not only is the situation inconsistent with the assumption that the hydrogen atom is a persisting material structure of radius approximately 10^{-8} cm, but moreover such an object could act as a source of energy that could be pumped indefinitely. As such, the object must be regarded as absurd on the broadest of physical grounds.

---

[61] Note that Sider himself is not a fan of grounding; this is simply him stating how the grounding theorist herself will put the matter.

Hence the 'inability to account for stable atoms in terms of classical trajectories of pointlike charged particles was the major problem of prequantum physics' – a problem of such magnitude that it prompted 'some serious people to question their existence – or at least to question the nice pictures drawn by chemists'.[62] As such, one of the original but fundamental successes of quantum mechanics was its recovery of the properties of this simplest of atoms.[63] And the way that the existence of the hydrogen atom was explained was by establishing that, according to the laws of quantum theory, a system of an electron and a proton comprises a bound state. It will be instructive to review how it does so.

As underlined earlier, central to the definition of a bound state is that its parts remain in spatial proximity to one another over time instead of separating to an arbitrary degree. This of course requires there to be a degree of attraction between the parts. Stated in energetic terms, the condition on a bound state is that the potential energy of the system arising as a result of their attraction exceeds the total kinetic energy of the parts. This in turn is to say, given the conventions made by physicists, that the total energy of the system is *negative*. Further, for a system to retain its structure over time and not collapse in on itself (and so not have its parts get arbitrarily close to one another), one furthermore requires that there is a *lower bound* on this negative energy.

The problem posed to quantum mechanics, then, was to show that the Schrödinger equation – the basic nomic template of quantum theory – when (a) fitted with the electrostatic potential appropriate to a $1/r^2$ force law (i.e. a potential of the form $V = \frac{1}{r}$) and (b) applied to a wavefunction composed of an electron and a proton admits of a well-defined and hence finite solution in which the energy ascribed to the system is negative. That is, one had to show that the time-independent Schrödinger equation

$$E\psi(r) = \frac{-\hbar^2}{2\mu}\nabla\psi(r) + V(r)\psi(r)$$

when supplemented with the electrostatic potential $V = -\frac{e^2}{4\pi\epsilon_0 r}$, yields an infimum.[64] It is now a standard exercise of early chapters of textbooks in quantum theory to show that this is indeed the case. And the explanation that we can give of why this bound state of quantum systems exists and yet not its

---

[62] Lieb (1990: 7).

[63] I say 'fundamental' because it is a success upon which much reductionist faith is built: see e.g. Hoefer (2003).

[64] Here μ is the 'reduced mass' of the electron–proton system defined by $\frac{1}{\frac{1}{m_e} + \frac{1}{m_p}}$.

classical analogue makes essential appeal to the quantum 'uncertainty principle'. As Elliot Lieb, one of the leading theorists of the stability of matter, puts it:

> The main contribution of quantum mechanics was to provide a quantitative theory that "explains" why the electron cannot fall into the nucleus. In brief, when the electron is close to the nucleus its kinetic energy–which could be zero classically–is forced to increase in such a way that the total energy goes to $+\infty$ as the average distance $|x|$ goes to zero. This property is known as the uncertainty principle (Lieb 1990: 7).

It should be noted here that the Heisenberg uncertainty principle, despite it being the most famous such principle, does not suffice for gaining quantitative estimates of the ground state energy. But there are many such principles, and the Coulombic form, Hardy form, or Sobolev expression of quantum uncertainty is typically that which is used instead. We need not however go into those details here. The point is that stability is rescued in quantum mechanics by a trade-off between the value of the total energy and the specificity of position; *how* these two aspects trade-off is governed in part by the form of the potential energy. Thus generically, given a set of particles, we can expect that the answer to the question of whether bound states will form will depend on the form of the potential describing the interactions between them, and hence on the form of the *laws of interaction* involved. In fact, whether bound states can form seems to depend on every feature of the laws one could think of.

To see this, note that in the classical framework, assuming that the $1/r^2$ form of the law persists to arbitrarily small distances, there is no bound state of electron and proton that can be formed. Thus whether a bound state forms depends on the *overall physical framework* – here, quantum or classical – within which laws are theorized. Further, we can see that whether a bound state forms depends on the *functional form* of the potential energy which, together with the kinetic energy, composes the total energy. Here we can note first that Jeans leaves it open that bound states could form were the form of the potential different at short distances. But we also know from quantum mechanics that stable bound states of the electron and proton would not form for arbitrary forms of the potential, for example, of the form $V = -\frac{1}{|x|^3}$. Indeed, one can show that a potential of the form $V = -\frac{1}{|x|^n}$ for *any* $n > 2$ does not allow for a lowest energy state, and hence does not give rise to well-defined physical objects at all.[65] Similarly, it is clear that whether a bound state forms depends on the *sign and strength* of the interaction. For if an electron and a proton repelled one another,

---

[65] This follows from the rescaling argument given in Lieb and Loss (2001: 271).

changing the 'minus' on the potential to a 'plus', then there would be no hope of forming a bound state; moreover, whether it does depends on *how* attractive it is. In sum, whether a bound state of two given particles can form or not depends on a multitude of features of the laws: the framework in which they are embedded, their functional form, and the sign and magnitude of the constants within them.

It seems to me, then, that a part – indeed a fundamentally important part – of what we ask physicists to do is to explain why the composite objects that we find around us exist: there is nothing a priori obvious about why they should exist and not others, as forming composites is not something we expect an arbitrary collection of objects to do. And the way that physicists will answer those questions is by showing that the laws of interaction among their parts permit the existence of bound states.

Here, then, is the issue for Humeanism. We ask scientists to explain why certain composites exist: this they will do in terms of the laws relevant to the parts of those composites. Such explanations connect different levels of objects, and as such qualify as 'metaphysical explanations'. (Remember: 'Just as causation links the world across time, grounding links the world across levels'.) The explanations of composites that metaphysicians offer typically give the grounds of a composite as its components plus the relation of binding between them. But it is this relation of binding that scientists aim to explain, and the way they do so is in terms of various features of the laws. It seems that laws, then, should be included as among the grounds of a composite object. But since composite objects in physics are identified with bound states – states that maintain *spatial proximity over time* – these explanations are *simultaneously* explanations of aspects of the structure of the manifold: not why it is that one event regularly follows another, but rather of why it is that certain confederations of properties are regularly found congregating in space. In explaining the levels structure of objects, then, we are *also* explaining certain aspects of the manifold. But we know what the laws are is determined, on the Humean view, by the totality of the manifold structure: the Humean says that they are 'grounded in', and hence 'metaphysically explained', in terms of that structure. Taken together, then, the Humean package looks circular.

In the previous incarnation of the argument – where the feature of the manifold that was to be explained was law-like temporal succession – a distinction between scientific *qua* causal explanation and metaphysical explanation could be made to break the explanatory circle. Here, however, the type of explanation at work is not happily characterized as 'causal', and is in fact well characterized as 'metaphysical'. For one thing, it is the time-*independent* Schrödinger equation that is doing all the work here, and this is not a law of temporal evolution; it is a law that describes stationary states of the system, hence in the first instance synchronic states of affairs.[66] But more importantly, the explanation tracks *levels of nature*, and as a level-connecting explanation it should be regarded as a 'metaphysical explanation'. How, then, if at all is the explanatory circle to be broken in this case?

A simple answer presents itself: and this is that we differentiate the types of metaphysical explanation involved. As such, the Humean should contend that *the relation of metaphysical explanation that explains laws in terms of patterns in the mosaic is not identical with the relation of metaphysical explanation that explains wholes in terms of their parts.* In particular, they should hold that the relation of metaphysical explanation that connects what we have called 'levels of metaphysics' is different from the relation of metaphysical explanation that connects 'levels of nature'. Given this distinction between the two types of explanations involved, we can say, as did Loewer, that the circularity objection 'falls apart'.

If this proposed distinction between forms of metaphysical implication holds up, then clearly the Humean is relieved of the charge of circularity, and as such the Humean metaphysics of nature retains whatever support it had before. But the distinction clearly also has a number of implications for the metaphysics of grounding and fundamentality. Perhaps the most obvious concerns the 'big-G grounding' debate. Recall that this is a debate over whether, when we make attributions of grounding in different philosophical contexts and with different relata, we are talking of 'fundamentally one and the same relation'. If this argument is right, then it seems we have to answer in the negative. After all, these two metaphysical explanations really must differ *qua explanations* if the circularity-avoiding strategy is to work: it cannot be that, as Audi (for example) contends, they differ solely in the nature of their relata. This calls into question that there is some unified and context-neutral notion of 'grounding' – a conclusion which has implications regarding the level of abstraction at which questions of the nature of grounding can usefully be addressed. Here we might note that while grounding pluralists often frame their worries by attacking the idea that the grounding relations encountered in wildly different philosophical contexts (semantic theory, ethical theory and so on) should be regarded as identical *qua* explanations, here it seems we can raise those worries while remaining squarely within the metaphysics of science. This suggests that theorizing about grounding may have to be even more tightly contextualized than pluralists already believe.

---

[66] Not every law describes evolution in time: Boyle's law – a stock example in the metaphysics of laws of nature – is one such example.

A second implication looks back to the argument given in the last section for the plurality of priority. Recall that in the last section it was mentioned that some metaphysicians regard themselves to be 'taking things a level deeper' than physics, given that metaphysicians are in the business of determining what ontological categories are fundamental. Were that the case, then what we took to be the fundamental level of nature ought to be regarded as 'splitting' into levels populated by members of different categories. But presumably metaphysicians can be said to be 'going deeper' than physicists only if they are engaged in a continuation of the same explanatory project. Given that – for the Humean at least – this is not the case, it seems we have further support for the sort of situation depicted in Figure 3 and hence for what I have called the 'plurality of priority'. While that in itself is of intrinsic interest, it in turn has implications for the methodology of fundamentality metaphysics. For if the hierarchies of being studied in metaphysics and science are different, then we will need to be very careful about drawing inferences about the former from features of the latter. (One consequence is that the conclusion of the next section framed in the context of levels of nature – namely, that we need not be committed to a fundamental level – may perhaps not have such radical implications for metaphysics as one might initially have thought or hoped.)

For these reasons, the argument given here for a multiplicity of relations of metaphysical explanation strikes me as of significant interest. But of course there are ways to push against it. For one thing, I haven't said much in support of the idea that there really are two distinct relations of metaphysical explanation, for I have pushed only that the Humean needs there to be. Nina Emery, for example, takes a similar argument to show that the Humean is in trouble, partly because there is no obvious 'independent reason' to regard the forms of metaphysical explanation as distinct (Emery 2019: 1554). Absent such a reason, anyone is free to *modus tollens* my *modus ponens* and argue that all that has been shown is that Humean metaphysics is circular and thus inadequate for that very reason. Furthermore, even if the Humean can use the above argument to make a principled distinction between forms of explanation, there is still a question of how, if at all, the argument can inform anti-Humeans, given that it is premised on a view of laws which they reject.

Let us take the first objection first. Meeting this objection requires some kind of criterion for the individuation of metaphysical explanations. I think it is fair to say that as things stand, there is not much literature to appeal to here – indeed this is an instance of the general point, made in Section 2, that the literature on metaphysical explanation in general is relatively underdeveloped.

However, Harjit Bhogal, in another work on Loewer's response to the circularity objection, offers a foothold for progress. For he argues, first, that the *aims* of explanations can serve to distinguish between types of explanation; second, that in general scientific explanation aims to 'unify' and that metaphysical explanation aims to 'elucidate the underlying structure of the world' (Bhogal forthcoming: Section 2.1). While I would take issue with this way of making the distinction, I think it is clear that the metaphysical explanations that are offered by science for why composite entities like the hydrogen atom exist do perform a unifying function, and also that this is a large part of why we value them. And I also agree that, however we characterize in positive terms the explanatory relation between the manifold and Humean laws, it is *not* an explanation that either achieves or aims at unification. (It is after all the supremely unifying description of the world – that given by the laws – that is itself being explained by a 'vast mosaic of particular facts', not vice versa.) This suggests that, for the Humean at least, there are positive reasons to differentiate the forms of explanation involved. *The conclusion is that Humeans should reject the unity of grounding.*

That argument of course went rather quickly, and there are places to push against it, But suppose for now that, for all the argument's sketchiness, that conclusion is correct. We can still ask what, if anything, those metaphysicians who do not subscribe to the Humean analysis should draw from the argument. Indeed, I myself am not a committed Humean: the debate is simply not one on which I have a settled view. So should *I* believe in the plurality of priority on the basis of my own argument?

The question here involves yet another methodological issue that probably deserves more attention than it currently gets. For in metaphysical disputes it is often asserted that it is a virtue of a metaphysical position that it fails to exclude 'live possibilities', even if the person asserting such a claim does not (one suspects) take those possibilities particularly seriously.[67] If we follow this precedent, then given that Humeanism is taken seriously by so many, even if not by me, it seems that I myself should actively deny the unity of grounding.

---

[67] An example of what I mean here may be found in Wilson, who as a 'primitivist' about fundamentality (see Section 1 above) objects to 'a characterization of the fundamental as that which is not dependent' on the grounds that it 'rules out various live accounts of what fundamental goings-on there might be, including self-dependent Gods, mutually dependent monads, and... partially-dependent strongly emergent features' (Wilson 2018a: 499). But one might well wonder whether there does not come a time when appeals to self-dependent Gods and monads no longer carry force in scientific metaphysics. And if that time is now, how can we appeal to these in an argument for the primitivist view?

However, while there is an intuitive sense in which this kind of generality should be seen as a virtue in metaphysics especially – given, once again, that metaphysics is often described as 'the most general' of the disciplines – I already flagged in the previous section that such permissivism risks giving rise to a certain insipidness. For me, then, given that I am not a Humean, the only way to cement the case against the unity of grounding is to go further than I have done so far, by rooting out compelling and relevant disanalogies between the two forms of explanation in play here.

Of course, I cannot here hope to do full justice to this question. And I concede that, on the face of it at least, there are reasons to worry that the argument will not generalize outside the Humean context. For crucial to the argument just given was the fact that the Humean does not take laws as primitive. As such, for Humeans laws need to be explained in terms of something fundamentally different in kind. But laws are so central to scientific theorizing in general that it becomes almost predictable that the level-connecting explanations offered by Humean metaphysics will be importantly different from those routinely deployed within science. Contrapositively, proponents of non-Humean, 'governing' views of laws have explicitly appealed to the similarities between the types of explanation that they rely on and those routinely used in science (see e.g. Armstrong 1983: 105). As such, there is reason to think that it is the Humean, and only the Humean, that has to recognize deep disanalogies with respect to level-connecting explanation.

Nevertheless, I think there is still room to challenge the unity of grounding even outside of the Humean context. For one thing, how contentful the analogy that Armstrong relies on has been powerfully challenged by Juha Saatsi, who reminds us that the power of inductive inferences is a highly contextual matter (Saatsi 2017). But I think there are some other disanalogies between the explanations offered in each context, and we should at least regard it as an open question as to whether these add up to a case for grounding pluralism. The most obvious is perhaps the fact that, while the contemporary explanation of the stability of hydrogen is canonized in textbooks and regarded as uncontroversial, there is no consensus among anti-Humeans as to how to understand laws (see Hildebrand 2020). This greater 'underdetermination' within metaphysics as compared to science is often attributed to its greater 'distance' from the deliverances of the laboratory. But to me it is worth reflecting upon whether it is something about the nature and the quality of the explanations offered by each field that partly accounts for the lack of consensus among metaphysicians as compared to scientists. Consider, for example, what Carl Hoefer has written about his encounter with the explanation of the hydrogenic bound state.

Working through the exact solution of Schrödinger's equation for the hydrogen atom was an important milestone in my formation as a fundamentalist.[68] I had never before been, and still was not, happy with [quantum mechanics] overall as a candidate fundamental theory; at a minimum, such theories should allow a coherent interpretation, and QM falls down badly on that front. Nevertheless, it offers us a well-defined differential equation and at least clearly says: "This mathematical law governs the structure of matter." When you work through the exact solution of the hydrogen atom, you see that in some very important sense, at least, this claim has to be right. The existence of a stable state, in which the proton and electron are bound to each other spatially yet never collapse as one would classically expect (and as one would also expect based on the ascription of their capacities qua oppositely charged things) falls out beautifully from the solutions of the equation....

What is particularly salient about the hydrogen solution is that its achievements transparently flow from the solution of an equation, and from nothing else.... From the 1930s onward, our understanding of hydrogen has been and will be based on a mathematical equation. (Hoefer 2003: 1404)

What Hoefer reports here is some of the *phenomenology* that accompanies the explanations given in science: one feels that one has learned so much, that somehow things will never seem otherwise. I close this section by asking the reader whether they have had a similar experience when reading metaphysics. If the phenomenology is indeed different across the two fields, then it is at least worth asking whether this is because there is a qualitative difference between the level-connecting explanations that are offered by each, and whether this is sufficient, given some purpose at least, for regarding the two forms distinct *qua* relations of explanation. But this will have to be work for another occasion.

4 Viciousness and Fundamentality

The previous section probed questions of the nature of grounding, and specifically that of whether it should be regarded as a unique and unambiguous notion. But questions about the structure of grounding, and thus about the order of levels that grounding determines, are just as replete with metaphysical and metametaphysical implications. It is the question of the well foundedness of grounding that will concern us here. This is the question of whether every sequence of grounding relations terminates in a first and itself ungrounded

---

[68] By 'fundamentalist', Hoefer means the view that there 'are truths, expressable in mathematical language, that accurately describe the behavior of all things in the physical world, at all times and places'.

member. If this is the case, then the existence of every non-fundamental entity is traceable back to a set of fundamental entities.

As we saw in Section 2, the idea that reality bottoms out into fundamental entities is often presented as part of the 'orthodoxy' surrounding the metaphysics of fundamentality. This doctrine, which we (as is standard) dub 'foundationalism', states that every chain of grounding relations necessarily terminates in a first and hence ungrounded member. It is the idea that 'that there must be a fundamental layer of reality, that is, that chains of ontological dependence must terminate: there cannot be turtles all the way down'(Cameron 2008: 1).[69] And while there are some illustrious defectors – perhaps most notably, Lewis, Armstrong, and Rosen – it probably is right to regard foundationalism as orthodox in this way.[70] 'For sure,' writes Westerhoff, 'anti-foundationalism can hardly be regarded as a majority view in metaphysics'. (Westerhoff 2018: 43). More strongly still, Cameron notes that

> [w]hether in metaphysics, epistemology, or ethics, Foundationalism has often been seen as the default, orthodox, view, with Coherentism [which allows for symmetric dependence among fundamental entities] being seen as the radical alternative. Infinitism [in which nothing qualifies as fundamental] is often simply dismissed, or not even considered as a live option (Cameron 2018: Section 2).

Foundationalism, then, is aptly regarded as a component of the grounding orthodoxy.

An obvious question to ask, then, is whether 'foundationalism' is true: must there exist a fundamental level, or is there scope for the more vertiginous anti-foundationalist view in which grounding chains tumble down ad infinitum? Given that, as was emphasized in Section 1, many contemporary practitioners

---

[69] It seems that Cameron's argument goes through just as well if we replace 'dependent' with 'grounded' entity. For the idea is 'that dependence cannot go on *ad infinitum*: there must be a fundamental level which grounds all the dependent objects or derivative facts (Cameron 2008: 8.)

[70] For example Rosen (2010: 116) writes that '[w]e should not assume that the [grounding] relation is well founded. That is a substantive question. It may be natural to suppose that every fact ultimately depends on an array of basic facts, which in turn depend on nothing. But it might turn out, for all we know, that the facts about atoms are grounded in facts about quarks and electrons, which are in turn grounded in facts about 'hyperquarks' and 'hyperelectrons', and so on *ad infinitum*. So we should leave it open that there might be an infinite chain of facts [p]←[q]←[r]←…'. Similarly Armstrong makes appeal to this same possibility in his argument for structural universals. He writes: 'a carbon atom consists of electrons, protons, and neutrons in a certain structure of bonding; protons and neutrons consist in turn of quarks; it is speculated that quarks in turn are composite…Maybe there is no end to this complexity. Maybe there are no simples, just structures of structures *ad infinitum*' (Armstrong 1978: 67–8)). It is this argument for structural universals that Lewis took to be the 'weightiest by far. Infinite complexity does seem, offhand, to be a genuine possibility' (Lewis 1986: 30).

conceptualize metaphysics as the study of the fundamental, this is would seem to be a question of supreme importance to the field. As Westerhoff puts it,

> [I]ts philosophical implications are substantial and have the potential to change the very way we think about metaphysics, and about philosophy more generally. For if there is no fundamental level of reality, what precisely are we doing in ontology and metaphysics, given that the search for such a level seems to be embedded in the very nature of these disciplines? (Westerhoff 2018, 43.)

Any assessment of how substantial those implications are, however, must bear in mind that there seem to be at least two distinct notions of 'levels' at play in contemporary metaphysical discourse (as was argued in the last section). And this question of a 'fundamental level' is typically addressed to what in our terminology would be called a 'level of nature': as I understand it, the question does not concern whether we should believe in a fundamental 'level of metaphysics' – that is, in the existence of a set of fundamental categories. Indeed, I know of no-one – certainly no realist about categories – who denies or so much as discusses this. But in any case, it will not be the metametaphysical implications of anti-fundamentalism so circumscribed that will be the focus here. Instead I will consider what attitude the naturalistic metaphysician should have toward the question of whether we ought to believe in a fundamental level in the first place.[71] This I will do by considering what the naturalist should say about the metaphysics and epistemology of infinite regresses.

Infinite regresses are inevitably central to this discussion because it is the idea that there something irredeemably problematic about *vicious* regresses in particular that buttresses the foundationalist world-view. To be sure, foundationalism 'seems to be more of an assumed *metaphysical axiom* (or *metaphysical law*) supported by intuition' than the product of any kind explicit argument (Bohn 2018: 169).[72] For few if any well worked-out arguments for foundationalism are to be found in the extant literature (cf. Bliss and Priest 2018: 17; Bohn 2018: 169). What one does find in the literature are repeated statements to the effect that without a fundamental level things would somehow 'never have got off the ground' (Cameron 2008: 6), or, more prosaically, that '[b]eing would be infinitely deferred, never achieved' (Schaffer 2010: 62). While too brief and gestural to be regardable as arguments themselves, their allusions to some version of 'infinite postponement' makes it 'not uncommon, nor unreasonable, to

---

[71] Here and going forward this should be understood to mean 'fundamental level of nature'.

[72] E. J. Lowe, for example, seems happy to simply state that 'an "axiom of foundation" [for ontology] is quite probably beyond conclusive proof and yet I find the vertiginous implications of its denial barely comprehensible' (Lowe 1998: 158; quoted in Westerhoff 2018: 27).

suppose that comments such as these can be reconstructed in the form of arguments from vicious infinite regress' (Bliss and Priest 2018: 18). But given the importance of the issue to metaphysics at large, a more explicit reconstruction of the connection from viciousness to the necessity of fundamentality is surely required.

Any attempt to articulate the argument from viciousness to foundationalism will have to provide answers to the following questions.

(Q1) What are the criteria definitive of a 'vicious infinite regress'?

(Q2) Do infinite regresses of grounds satisfy those criteria?

(Q3) Does satisfaction of those criteria entail some kind of metaphysical contradiction?

Of these, (Q3) is probably the most involved. For while it seems obvious that 'viciousness' – a normatively loaded term – constitutes something 'bad' and as such something one might want to avoid in one's theorizing, it seems often to be simply taken for granted that that means that *the world itself* will not sanction viciousness. To my mind there is – as with almost any move from 'theoretical virtue' to claims about the world – nothing remotely obvious about this inference. Nor is it obvious to me, however, that vicious infinite regresses of grounds *are* metaphysically possible. The denier of foundationalism, then, must show either that the criteria of viciousness involve no such contradiction, or that consideration of the issue is simply moot. The question that serves as our focus here is what resources the naturalistic metaphysician can offer to an attempted resolution of this issue.

While I do not affect here to fully resolve the debate, I will make two claims that at least advance it. The first claim, which will draw heavily on recent work by Ricki Bliss, is that there is no reason why an infinite regress of grounds *described in the language of science* must necessarily be guilty of viciousness. Rather, viciousness may simply be an artefact of conceiving of a descriptively rich situation in too descriptively austere terms. If that is right, it follows that the naturalistic metaphysician – the metaphysician interpreting the world *as given to us by science* – can happily sanction infinite regresses, and can moreover do so without even bothering to engage with the question of whether vicious regresses are possible. Such a finding means that (Q3) is moot – surely a mark against the foundationalist orthodoxy. But it turns out that we can go even further; for I will then claim that any infinite regress of grounds *known through the methods of science* likely always *will* exhibit a form of viciousness. The conclusion the naturalist will want to draw from this is not only that foundationalism is false, but that even those regresses we regard a priori as especially problematic ought to be sanctioned in metaphysics – answering (Q3) in the negative. And while there will certainly be less naturalistic metaphysicians who will want to reject that conclusion, doing so, I will argue, is costly.[73]

4.1 What Makes an Infinite Regress 'Vicious'?

Arguments from vicious regress are not only used to support metaphysical foundationalism but are rather something of a 'mainstay in the philosophical tradition' (Bliss 2013: 399). Like virtually everything in that tradition, however, there is disagreement about what makes a regress classifiable as 'vicious'. One thing that there is agreement on is that not every infinite regress one encounters in philosophy qualifies as pathological, for some are regarded as 'benign'. A foremost example of regress universally regarded as benign is the Tarskian truth regress, in which every truth *P* entails the further truth that *P is true*, and so on ad infinitum. By contrast, Bradley's regress, the 'turtle regress', and the regress embedded in the homuncular theory of perception are all regarded as paradigms of the vicious. By what criteria, we should ask, is a regress to be categorized as one or the other?

While several diagnoses of what makes the difference have been offered – including that they harbour contradictions, unduly violate parsimony, or are in conflict with a 'known finite domain' – Bliss argues that the crux of the distinction lies in whether the regress concerned exhibits a certain kind of explanatory failure.[74] This makes it sound like grounding regresses are always necessarily vicious, and one novelty of Bliss' presentation is that it shows that, relative to the 'explanatory failure' criterion, this need not always be the case. As she puts it, these paradigmatic vicious regresses

> are generated as a consequence of continually attempting to overcome an explanatory failure that arises at the first level of the analysis. The problem [...] arises at the first level of analysis, and continues to recur at each level...

> The explanatory failure that occurs at the first level of the analysis consists in the fact that the *explanans* is of the same form as the *explanandum*: the phenomenon for which we are seeking an explanation reappears as its own explanation (Bliss 2013: 410).

---

[73] I should note that Morganti (2015) also argues that the squeamishness with which philosophers regard infinite regresses is unwarranted. Our arguments are however independent. My argument, for example, uses facts about the quality of explanations to defend the possibility of infinitism, while Morganti takes it to be a virtue of his position that it is 'also acceptable for those who explicitly refuse, for whatever reason, to establish a tight connection between metaphysical structure and explanatory structure' (Morganti 2015: 564).

[74] To be clear, Bliss is not the first to define viciousness in this (rather intuitive) way: rather, she defends an analysis that has been offered before (e.g. in Passmore 1961) from later criticisms. I also note that other recent writers have claimed that the distinction resides in the direction in which the series is 'closed' relative to the direction of explanatory dependence (Maurin [2007]).

As an example of this phenomenon we can cite the Bradley regress.[75] In simple terms, the Bradley regress occurs when we seek to explain how it is that an object instantiates a property; or, for maximal lucidity, how it is that an object instantiates a relation, such as the Eiffel Tower's *being taller than* Big Ben. We are told by the Platonist that it does so by standing in a relation of instantiation to the relation *being taller than*. But then it seems we have a new question of how it is that the Eiffel Tower instantiates *that* relation to the relation we were interested in the first place. This seems a clear case in which there is an explanatory failure at the first stage insofar as the explanans has the 'same form' as the explanandum. For each explanation has the form of 'the object stands in a relation', which is what we sought an explanation of in the first place: the particular relations involved may be different, but it is not these particularities we are interested in. We are interested in why *objects instantiate relations*, and the answer one gets is of exactly that form.

Something along these lines does seem to capture what is wrong with (what we are apt to regard as) other paradigm vicious regresses. For insofar as the homunculus theory posits a further person to explain why an initial person has experiences, the *very same question* arises with regard to *them* as to how *they too* have experiences (even if those experiences are different); insofar as a further turtle is posited to explain why an initial turtle stays suspended in space, the *very same question* arises with regard to *them* as to how *they too* are suspended (even if they are suspended elsewhere in space). In each case, '[each] member of the regress is numerically distinct from that which precedes it; *qua explanans and explanandum*, however, they are identical. And in so being, we never break out of the explanatory failure that gives rise to the regress by breaking into the cycle with adequate explanation' (Bliss 2013: 412).

Given its intuitive correctness and seeming extensional adequacy there is a strong case to be made that the diagnosis Bliss offers is on the right track. But the devil is in the details, and in particular more needs to be said about what is meant by the crucial notion of 'sameness of form' that lies at the heart of the analysis. Bliss makes no claim to having an analysis of this notion, and instead offers only some useful pointers. The first is that it does not mean merely sameness of *logical* form. For example, she claims that we might explain *my existence* (as the qualitatively complex living thing that I am) in terms of my organs' functioning; my organs' functioning in terms of their molecular functioning; each of these in terms of their atomic functioning; and so on. In *this* case, she says, 'our explanans and explanandum are *not* of the same form', and any ensuing regress may accordingly be regarded as benign – even though the

---

[75] At least on one interpretation of it: see Cameron (2018).

explanation offered in each case is of the same logical form as the explanandum (Bliss 2013: 414). This example might be taken to suggest that if the *kinds* involved at each stage are different then the explanatory failure can be avoided. But this can't be right, as we can see in the case of the Bradley regress; furthermore, as she points out, it's not as if the turtle regress would be greatly improved were it modified to involve different species at each stage. As she notes, 'what is relevant [is] the explanatory role such zoological exotica are playing *qua* objects of support [and] not whether they are the same in other respects. The version of the regress that involves only turtles just wears its viciousness a bit more boldly on its sleeve' (Bliss 2013: 412).

What we begin to see, then, is that the differences relevant to maintaining the 'same form' of explanation is not simply a logical but in part also a contextual affair: a regress of grounds may or may not be vicious depending on what it is that we are interested in explaining. And of prime importance for our purposes is that the *level of abstraction* with which we apprehend our explanatory target may flip a benign regress into a vicious one. For as Bliss points out, if we are not interested in explaining me in all my variegated living glory but rather the mere fact of my existence, then the explanation offered in terms of my organs, their composing molecules etc. plausibly *is* guilty of explanatory failure given that each of those must be presumed to exist in order to discharge their explanatory obligations. As she puts it, '[t]he regress is not benign, however, if what we are seeking an explanation for is how anything exists, or has being, at all' (Bliss 2013: 414).

Although Bliss' proposal is not yet fully developed there are important lessons buried within the dynamic she highlights. For as with Wilson's response to the 'what grounds Grounding?' question, what it shows yet again is that there is something about the level of abstraction associated with metaphysics – something intimately connected to its aspirations to maximum generality – that makes it so vulnerable to asking questions that cannot be answered satisfactorily. And while the idea that metaphysics routinely asks questions that we cannot satisfactorily answer is a theme long familiar from Kant, here it seems solely the nature of the question we are asking that is generating all the problems and not anything about our relationship to it. Perhaps, then, the key to avoiding viciousness in worlds of infinite descent is to ask the question differently, and in the more discriminating terms of science – terms that view a molecule as very different from an organ, and from a person, and from an atom. For I think we should agree with Bliss that insofar as we are interested in explanations of worldly entities conceived of in the less abstract language of science – such as persons in all their variated glory – then in the regress above we *do* have a satisfactory explanation at each stage.[76] And further – looking back to the previous section – we can rightly regard these explanations as *metaphysical* explanations (being 'level-connecting', non-causal and so on). If each of those is right, then the naturalist who wants to deny foundationalism need not even roll up their sleeves and confront (Q3) in order to do so legitimately, as the question has been shown to be moot. A world described in the descriptively richer terms of science can be one in which there are infinite grounding chains that do not exhibit viciousness. So if it is viciousness that is supposed to make us suspicious of such worlds, there is no reason to regard them as impossible.

To be sure, the foundationalist will no doubt object at this point that 'upping' the level of description merely serves to paper over the true metaphysical problem, which is that the explanations for the *existence* of these entities ultimately remain as vicious as ever. But it seems to me that since – channelling some amateur Kant again – any entity has to have properties in addition to existence in order to exist at all, *the benevolence implicit in the less abstract regress should be taken to trump the viciousness implicit in the more.* More of course more would need to be said in a full defence of that (big) claim. But it nevertheless seems clear that there is a similarity between the situation here and Wilson's objections to 'big-G grounding', in that in both cases we see how transgressing to too high a level of abstraction can produce trivializing or otherwise pathological results. In both cases, it seems perfectly reasonable to maintain that the solution is simply to reinstate the required descriptive detail. It thus seems to me quite open to the naturalist to maintain, on all sorts of principled grounds, that the problem lies more with the nature of the questions we choose to ask about them rather than with infinite regresses in themselves.

In sum, it seems we can say the following. Infinite regresses of grounds need not be vicious; whether they are is dependent on whether the 'same form' of explanation is given at each stage. And whether that is the case is a function of our explanatory interests, and in particular the degree of abstraction with which we apprehend the explanandum. While I'll concede that more needs to be said regarding the nature of 'sameness of form', for now we can rest with a case-by-case judgment, and so far that judgment delivers no reason to think that the metaphysical explanations that are furnished *by science*, expressed in its more discriminating language and given its less abstracted interests, necessarily exhibit sameness of form. As such, its regresses are not 'vicious'. Taking

---

[76] As she later put it at the 2019 Pacific division meeting of the American Philosophical Association, 'I don't believe that we would deny that we have done a whole lot of explaining here. Indeed, whole disciplines are built around discovering and understanding explanatory patterns at the non-fundamental level'.

a naturalistic approach to grounding, then, and hence focusing on the metaphysical explanations that are furnished by science, would seems to create an environment more hospitable to infinite descent than that assumed by the foundationalist.

That concludes the first part of what I want to say about how taking a naturalistic perspective on grounding interacts with arguments from 'viciousness'. What I want to argue now, however, is that things are not quite so simple as they have been made to seem so far: for while there is something about the *language* of science and thus the terms in which it frames explanations that makes them less susceptible to accusations of viciousness, there is something about its *epistemology* that makes vicious regresses the only infinite regresses that it can establish as real. Thus while there is no reason why any given instance of metaphysical explanation offered by science will necessarily exhibit 'sameness of form', it seems those that feature in any *infinite* chain of explanation whose existence is justified by science necessarily will. This of course complicates the picture. But with a sufficiently robust commitment to naturalism it only strengthens the case against foundationalism.

4.2 The Empirical Justification of Infinite Regress

The claim I want to make now is that the infinite regress of grounds that are scientifically sanctioned by science are always liable to be vicious. First on the agenda, then, is to consider how, if at all, we can justify our belief in infinite grounding chains by empirical means. An obvious starting point for this discussion is Schaffer's influential paper of 2003, *Is there a fundamental level?*, which of course we met in Section 3. At that time, as he noted, the issue of whether or not the levels hierarchy suggested by science bottoms out into fundamentalia was 'almost entirely neglected' (Schaffer 2003: 498) – something that cannot be said now, and arguably largely as a result of this paper.[77] However, I will argue that this paper at best demonstrates how *not* to use empirical considerations to argue against the existence of a fundamental level.

Schaffer's paper aims to challenge the idea that there exists a fundamental level to the actual world on empirical grounds. As such, it challenges what is called 'fundamentalism'. Clearly such an argument is simultaneously an attack on 'foundationalism', as the latter asserts that there exists a fundamental level to every world whatsoever. What was novel about this paper was the fact that it

---

[77] To be clear, Schaffer is not the first person in recent memory to propose the coherence of a world lacking in foundations, and to incorporate its implications into metaphysical schemes more broadly. As it happens, it was the dynamic between Armstrong and Lewis, cited above in footnote 70, that inspired my own research on the topic.

opted not to approach fundamentalism in an a priori way. For as far as Schaffer is concerned, just as the existence of the 'hierarchy of nature' is a delivery of empirical science, the question of whether this hierarchy is well founded should likewise be construed as empirical. As he puts it:

> Since the fundamentalist begins with the hierarchical picture of nature, which is an empirical thesis based on the idea that the structure and discoveries of science reflects the structure of nature, the thematic fundamentalist ought then to look to the structure of science for empirical evidence for a fundamental atomic structure in nature....
>
> [I]f fundamentality is to have bite, it must be an empirical discovery, rather than a mere methodological presumption (Schaffer 2003: 501–2).

As per his belief, alluded to above in Section 3, that compositional structure is the 'central connotation' of levels structure, he expresses fundamentalism in terms of atomism here. In what follows, I suggest we be as liberal as possible in interpreting 'atomism', and so understand it simply as the doctrine that there exist fundamental entities that in some sense account for, and in that sense ground, non-fundamental entities of broadly the same category (it being 'levels of nature' that we are concerned with after all).

The main takeaway of the paper is that 'the empirical evidence to date is neutral as to which structure science is reflecting' (Schaffer 2003: 505). This is supposed to come as a surprise, as the general presumption is in favour of fundamentalism (foundationalism being 'orthodox' after all). Now, I should admit that it is actually not clear to me from reading this paper what Schaffer takes the empirical evidence favouring fundamentalism to be.[78] But here in any case I want to focus on the evidence provided for the other side of the ledger – evidence that is deemed sufficiently strong to balance whatever it is that provides support to fundamentalism. The argument is inductive in character and, in a nutshell, it is this.

> The history of science is a history of seeking ever-deeper structure. We have gone from 'the elements' to 'the atoms' to the subatomic electrons, protons and neutrons, to the zoo of 'elementary particles', to thinking that the hadrons are built out of quarks [...] Should one not expect the future to be like the past? (Schaffer 2003: 503).

The structure of the argument here is a 'meta-induction', to use language that will be familiar to philosophers of science – that is, an induction on the history of science. Based as it is on historical facts there is thus at least some sense in

---

[78] He does little more than gesture at Weinberg's famous quote regarding the 'convergence of the arrows of explanation, like the convergence of meridians toward the North Pole' (Schaffer 2003: 503). But that evidence is hardly in conflict with the history of science which forms the basis of his anti-fundamentalist meta-induction, to be outlined momentarily.

which it is an empirical argument of sorts. Moreover, some naturalistic philosophers do regard it as a compelling inductive argument against fundamentalism. Here for example is Ladyman and Ross:

> Arguably we do have inductive grounds for denying that there is a fundamental level since every time one has been posited, it has turned out not to be fundamental after all. (Ladyman and Ross 2007: 178).[79]

However – and regardless of the strength of the argument for the contrary conclusion – I would argue that this inductive argument does not succeed in establishing anything whatsoever about the well-foundedness of the levels structure of the world. Thus it seems to me that if one wants to argue against the existence of a fundamental level on empirical grounds one has to do something different.

There is a lot that one could say in defence of this claim, but let me confine myself just to a couple of remarks that have not yet made it into the literature.[80] Some have already objected that it is surely stretching the inductive evidence – a handful of cases – beyond breaking point to take it to support the in-principle infinite amount of work that the argument needs it to do. It is, in Callender's words, an 'unwarranted inductive inference of the wildest sort, comparable to Schaffer asking me for chocolates from a box I have, receiving them 5 times, and then concluding that I must have an infinity of chocolates' (Callender 2001: 3). Sider similarly mocks essentially the same argument, comparing it 'to the argument that there must be infinitely many people, since for each person we've observed there exists a taller person' (Sider 2011: 135). Of course, it isn't difficult to fault the inference from a few cases to the existence of a further infinity in almost any context, least of all this: after all, if there is a fundamental level to reality we can expect it to take us few shots to find it! But in my view, each of these comparisons is nevertheless a little unfair. For, in each case, we have *background knowledge*, about chocolates and people respectively, that tells us the induction is ridiculous. We do not, however, have similar background knowledge about Schaffer's case, and so we cannot straightforwardly infer that this induction is just as bad.

The true problem with the argument, to my mind, is not that we possess background knowledge that reveals that the induction is a bad one, but rather that we lack the background knowledge needed to make any normative assessment of it whatsoever. For as is well known, in assessing the strength of an inductive inference it never suffices to attend only to the form of the inference involved. Paradigmatically in the case of enumerative inductions, one needs information

---

[79] See also Bohn (2018: 177) for a similar endorsement.
[80] See McKenzie 2011, Section 2 for further criticisms.

about the *kinds* of entities involved, and the general rule of thumb is that the closer the kind involved approximates a 'natural kind' then the safer the induction.[81] (If this isn't immediately obvious, consider that while it may be safe infer from the fact that all samples of potassium in your lab have melted at X degrees Kelvin that all samples of potassium do, it is presumably not safe to infer from the fact that all the wax in your lab has melted at X degrees Kelvin that all samples of wax whatsover will.) *Here*, however, it is not remotely clear that we are dealing with a 'natural kind' at all.[82] If anything, we are dealing with a sort of 'supra-kind' that incorporates all natural kinds that could ever be studied in science, and we simply do not know what sort of rules would govern inferences with it. Certainly, there is no obviously *scientific* knowledge for us to draw on in assessing the reliability of the argument: the vast majority of the particles involved do not feature in known theories after all. It thus seems that if there are any rules governing the legitimacy of this inference they must derive from a highly abstract and largely a priori theory – in other words, from something like a priori metaphysics. But this argument against fundamentality was supposed to be novel in that it is rooted in empirical facts. As such, all we can say is that maybe this induction is as ridiculous as the chocolate box inference; but maybe it is not. We simply have no means at our disposal to settle this.

For these and other reasons, I hold that there is no defensible meta-inductive argument based on the history of science either for or against fundamentality.[83] If we are going to deny fundamentality on naturalistic grounds, then, another strategy is required. Given the seeming hopelessness of trying to look around the corners of future science and predict what it will do, it seems the best we can do is motivate anti-fundamentalism *from within the perspective of an extant physical theory*.[84] That is, *we should believe in anti-fundamentalism if, and only if, we are confident in a physical theory that has it among its implications.* This I call the 'internal' approach to anti-fundamentalism.[85] While that in itself might seem naturalistically well-motivated – it is, after all, simply an extension of how naturalists typically justify other claims in the metaphysics of science – it remains an empty *a priori* prescription until we can establish that it is possible for a physical theory to actually do so. But it is not at all obvious that it can. Does

---

[81] Such contextuality is the basis of Norton's 'material theory' of induction (Norton 2014).

[82] Bliss and Priest (2018, 25) make a similar remark (though in a slightly difference context): 'There are going to be difficult issues associated with the thought that "dependent entity" and "fundamental entity" are kind-terms. Where "dog" seems like a good example of a kind term, it is less clear that "dependent entity" is'.

[83] I discuss other objections in McKenzie 2011, Section 2.

[84] On the futility of making predictions about future science, see Rescher 1983.

[85] McKenzie 2011.

not any theory have to have to start somewhere, and would that not commit us to fundamental ontology of some sort?

The answer here is: yes and no. Yes, because all theories must of necessity start somewhere. But also no, because the place that they start does not need to be with fundamental *ontology*. A real-life example of how this can happen may be found in the S-matrix theory that was popular in high-energy physics in the 1960s. This theory was the precursor to the current, quantum field theoretic theory of the strong nuclear interaction, quantum chromodynamics, that represents one of our greatest scientific success stories; it was also the incubator of string theory.[86] While ultimately falsified, this was a theory that was taken to imply that there were *no fundamental strongly-interacting particles*, or 'hadrons' – the family of particles studied in nuclear physics and that includes the proton, neutron, and pion. Rather, every token hadron was regarded as a bound state, but not yet of quarks: rather, every hadron was ultimately taken to contain hadrons of every type – including further tokens of its own type. As such, it described a form of 'gunk world' in the language of metaphysics: a world in which there may be fundamental laws, and fundamental properties, but no fundamental objects. Since this is a theory I have discussed in detail elsewhere, and since the full argument is rather complicated, I will here provide only a schematic outline of the way in which S-matrix theory has anti-fundamentalist implications, and as such provides us with 'proof of concept'.[87]

- What occasioned the introduction of the S-matrix theory was the realization any equation – any Lagrangian – describing the strong nuclear interaction was, on account of the strength of the interaction, too strong to be amenable to the standard methods of solving problems in particle physics (namely, perturbative methods). In that respect, those equations would be unsolvable and non-predictive. A new approach was needed if these interactions were to be described.
- The approach taken by S-matrix theorists was to try to compute the probabilities of interaction outcomes directly from some assumed general principles governing the dynamics. These included the assumptions of Lorentz invariance (to make the theory relativistic), unitarity (to make it quantum mechanical), plus an assumption that the probability functions were analytic functions of their variables.
- The scattering matrix, or S-matrix, is a mathematical object providing the probabilities that any one particle or set of particles will be produced from

---

[86] See Cappelli et al. (2012) for a review of the historical and conceptual connection between these two theories.

[87] For a fuller account, see McKenzie 2011.

'scattering' others (that is, colliding them). In particular, the elements S_{ij} of this matrix give the probability that a set of particles $P_j = \{p_1 \ldots p_m\}$ can be produced by scattering a set of particles $P_i = \{p'_1 \ldots p'_n\}$ off one another. Given the assumed symmetries of the strong interaction, each output state must have the same *quantum numbers* as the input (same total value of spin, electric charge, strong isospin, etc).

- Sometimes single intermediate particles with the same quantum numbers as the input can be produced as a 'bound state' via scattering processes. As per the discussion of the previous section, anything producible in such a way was unambiguously regarded as composite and hence non-fundamental. The properties of any composite particle cannot be arbitrarily assignable, or in other words regarded as 'free parameters'. Rather they will be derivable from – since 'grounded in' – those of the particles involved in producing it. Thus suppose that a particle C_1 was created from colliding particles p_1 and p_2, and C_2 from particles p_3 and p_4. Then the properties of C_1 will be a function of those of p_1 and p_2, and C_2's a function of those of p_3 and p_4.

- The S-matrix was highly 'holistic', in that the probability of transitioning from P_i to P_j, encoded in S_{ij}, and thus the probability of producing a composite particle with the same quantum numbers as the P_i, turns out to ultimately be a function of *every particle in the theory*. As Geoffrey Chew, the chief architect of the theory, was to put it, 'No [composite particle] can be completely understood without an understanding of all the others' (Chew 1968: 765). It follows that the properties of the composite C_1 above are functions of those of C_2, and hence of p_3 and p_4, in addition to p_1 and p_2. This means that the properties of the composing particles p_1 and p_2 are *correlated* with those of p_3 and p_4.

- That lack of independence among the properties of all strongly interacting particles led practitioners of this theory to suggest that the dynamics of any one interaction were so bound up with those of others – and vice versa – that there may in fact be no free parameters in the theory whatsoever. Rather, every property of every particle was a function of those of all the others. As Chew put it: 'In this circular and violently non-linear situation it is possible to imagine that no free parameters appear and that the only self-consistent set of particles is the one we find in nature' (Chew 1964: 34).

- Since to be composite is to have properties that are not arbitrarily assignable, it was then hypothesized that *every strongly-interacting particle is a composite*: a composite of any collection of particles with the same quantum numbers. This idea that every hadron is a composite of all the other hadrons became known as 'nuclear democracy'. (In Chew's words, '[i]f one wishes to relate this idea of particle democracy to the older language of bound states or composite particles, it amounts to saying that each particle is a composite of all the others'.) A token instance of any kind of hadron was then taken to contain as parts tokens of every other kind of hadron, including those of its own type, and so on *ad infinitum*.

- Together with the 'analyticity' postulate governing the dynamics, this assumption of no free parameters also implied, it turned out, that the S-matrix was a linear function of the spin, s, for all possible values of spin $s = 0, 1/2, 1, 3/2. . . . . .$

- This linearity was demonstrated for all but the lowest values. However, that linearity held for spins $= 0, 1/2, 1$ remained a postulate.

- The assumption of linearity for all values of spin turned out to have a clear empirical signature (the 'Regge trajectory'). As such, there was a means of 'testing the idea of nuclear democracy that there are no elementary particles'.[88]

- After some initial predictive successes, the theory was eventually was falsified under testing: the singularity of the pion (with spin $= 0$), for example, failed to exhibit the required linear behaviour.

While that was of course compressed and schematic, the hope is that it at least gives a glimpse of how the S-matrix theory was able to imply a world of infinite descent. The highly holistic structure of the theory was taken to point to the existence of a 'gunk world' in which every type of particle in its ontology was a composite of tokens of every type; that postulate was then equated with a condition on the form of the dynamics, which was in turn translated into a clear empirical hypothesis (and subsequently falsified). As such, I see no reason why we cannot view this as an example of a real physical theory that can be reasonably interpreted as having anti-fundamentalist implications. Now of course, given that the theory *was* falsified, I make no claim that it has any implications about the actuality of infinite descent. However, it seems to me it does at least give 'proof of concept' of the idea that arguments against fundamentality need not be meta-inductions: rather, it can be the internal logic of a physical theory – the implications of its system of postulates – that can be used to furnish a denial of the existence of fundamental entities. As such, while I see no hope, for the reasons I have given, for meta-inductive arguments against fundamentality, there *is* positive reason to believe that internal arguments against fundamentality may be furnished by the best of our future theories (whatever they may turn out to be). And if that plays out, we will have supreme naturalistic reason for denying the existence of a fundamental level.

---

[88] Regge trajectories are later identified with the string tension associated with a particle in string theory. This is only one of the ways that S-matrix theory served as an incubator for that theory.

What I want to argue now, however, is that while internal arguments offer the most promising means of denying fundamentality – and hence for rejecting foundationalism – they are in tension with the idea that the only acceptable grounding regresses are non-vicious. As such, internal arguments present us with new resources with which to confront (Q3).

4.3 The Tension

To see the issue here, recall the comment above that the answer to the question of whether a physical theory must commit us to *something* fundamental is always 'yes and no'. For example, we know that it is true that there is no particle – no hadron kind – that is regarded as fundamental according to the S-matrix theory. But the very concept of an internal argument against fundamentality presupposes that the anti-fundamentalist must in all cases be committed to something that is at least *methodologically* fundamental: the set of basic physical postulates from which their ontologically anti-fundamentalist conclusions follow. (In fact, Chew himself was very open about the fact that his key 'analyticity' postulate in particular was simply taken as axiomatic, issuing from no more fundamental source.) This has the consequence that internal arguments against fundamentality are limited in a crucially important sense: by definition, they proceed from a set of postulates, formulated by means of a finite set of predicates. As such, it seems that only a certain, circumscribed amount of qualitative variation is permitted in the descending ontologies they are capable of fully describing. The picture that S-matrix theory presents us with, for example, is one in which compositional chains go on forever, but the types of particles that feature in these chains recur *ad infinitum*. Thus although Schaffer does not define the term, it seems clear that this would be a prime example of what he (and now others) calls a 'boring' world – one in which the property structure repeats itself indefinitely as we plunge deeper down chains of determination. The example that he gives of such a world is Pascal's:

> Blaise Pascal, for instance, believed that each part of matter housed a micro-universe with a miniature earth, sun, and planets, and that each part of matter of this micro-verse housed a micro-microverse, ad infinitum. On Pascal's worlds-within-worlds picture there is of course no fundamental level of nature, but there presumably could be a complete physical theory applicable at every level, as long as the same dynamics applies throughout (Schaffer 2003: 505).

With a suitably liberal reading of 'same dynamics', I think that any internal argument against fundamentality will be limited in at least something like this way. Theories, after all, can only contain so many predicates. As such, although the degree of homogeneity in the descending sequence need not in general be so dramatic as it is in the S-matrix theory, it seems the only kind of anti-fundamentality we can hope to establish on naturalistic grounds is a sort of 'half-way house' in which the overall framework and its stock of predicates stays the same even as the determination structure never ends.[89] That this is a less radical form of anti-fundamentalism than could be imagined should be immediately clear. Nevertheless, it is a form of anti-fundamentalism, and one that the sciences have a shot at licensing.

This consequence surely has implications of some sort for the standing of vicious regresses in metaphysics. Recall that it is widely accepted in metaphysics that while not all regresses are problematic, vicious regresses are sufficiently pathological to preclude their reality. And recall that vicious regresses are plausibly characterized in terms of a certain explanatory failure: namely, the recurrence of an explanation of the 'same form' at each level. While of course this all needs further articulation, it seems clear that insofar as internal arguments commit us to (something like) 'boring worlds', the regresses involved will by this definition *inevitably be guilty of viciousness*, of some degree at least. Indeed, Schaffer himself compares boring worlds to the turtle hierarchy, which of course is a paradigm of a vicious regress. And we have seen how this plays out in the S-matrix theory, insofar as the metaphysical explanation of why there is a hadron of a certain kind – a neutron, say – is inevitably going to make reference to the existence of another hadron of that same kind and of which it is at some level composed. So while a full defence of this claim would need both a better sense of 'same form' of explanation and a more general characterization of the sorts of viciousness that are, and are not, inevitably consequences of 'internal arguments', given that the form that scientific explanations take is inevitably constrained by the basic postulates of the relevant theory there is every reason to think that the resulting regresses will be vicious in some substantive sense. And I underline that that this will be the case even when these explanations are framed in the language of the science itself: the viciousness involved in the metaphysical explanation of a neutron in S-matrix theory, for example, is not a pathology incurred by 'upping the abstraction'.

It seems, then, that while the grounding explanations offered to us by science need not be guilty of viciousness, those *infinite* chains of explanations that science can give us any reason to believe in necessarily will. What, then, are the morals of this for the standing of vicious infinite regresses in metaphysics, and thus for our theories of grounding and fundamentality?

---

[89] Schaffer calls it 'supervenience-only fundamentality', and distinguishes it from 'full-blown fundamentality'.

It is clear what the naturalist is going to *want* to say at least. They will say that the possibility of such regresses has been sanctioned by a real scientific theory, and such theories are as good a guide to reality as we are going to get. If we have a real scientific theory that conflicts with foundationalism, then, the only conclusion to draw is that the latter has been shown to be false. To twist the knife further, they might note that taking foundationalism seriously would have resulted in the abandonment of research in S-matrix theory as soon as it was realized that it had infinite regresses among its consequences. Not only would this have been an inappropriate intrusion into the epistemology of science at a time when it was still regarded as an open empirical possibility, it is one that could well have derailed the programme before it had paved the way for the emergence of string theory – that which is, according to many, the best candidate for a truly fundamental theory. And while any given analytic metaphysician might hold that metaphysics is 'autonomous from' or somehow 'prior to' science, presumably none would want to be charged with actively frustrating scientific progress in this way.

I myself am on the record for drawing such critical conclusions from this case study. And we need to be clear that if this is the right conclusion it is surely a significant one. Many, after all, are deeply convinced of the truth of foundationalism – it is the orthodoxy after all – and as the quote from Westerhoff above reminds us it is surely consequential if it is false. Furthermore, the sort of infinite regress that is here sanctioned has been regarded as more problematic than the sort of 'nucleus $\rightarrow$ protons $\rightarrow$ quarks $\rightarrow$ ...' regresses that may permit infinite qualitative variation and hence potentially not exhibit any viciousness. But for all that – indeed largely because of that – we can only expect there to be pushback; and there are certainly places to push. The most obvious criticism to make of the argument rests on the fact that the S-matrix theory, while in some sense a 'real scientific theory', is nevertheless unambiguously a false one. This leaves it open to the committed defender of foundationalism to say that it is hardly surprising that the S-matrix theory is false given that (by their lights) it is not even metaphysically possible, however much an open *epistemic* possibility it may have been taken to represent at the time. Nor will they likely be moved by the heuristic argument just made against them either. There are after all already arguments in the literature that assuming a fundamental level inhibits scientific imagination, ambition, and progress.[90] If they were

---

[90] David Bohm, for example, adduces this view in his famous discussion of the 'qualitative infinity of nature'. Here he asserts that 'not only can nothing of real value for scientific work be lost if we adopt the notion of the qualitative infinity of nature..., but on the contrary, much can be gained by doing this' (Bohm 1957: 136.)

going to be moved by such arguments they would have been so already, without this convoluted detour through S-matrix theory. Thus it seems that only if we have a *true* scientific theory that supplies an internal argument against fundamentality will the committed foundationalist budge. And that we still do not have.

If this is the line the a priori metaphysician wishes to take then it is a consistent one at least. But note how much work in metaphysics it shuts down if applied consistently. It in effect reduces metaphysical possibility to physical possibility, to what is sanctioned by the models of a true physical theory. Ironically this is an even more restricted sense of metaphysical possibility than seems to be recognized, implicitly at least, by most naturalistic metaphysicians. For this reason, it seems to me the a priori metaphysician likely does far less violence to themselves by taking the above example to undermine foundationalism instead of their working assumptions regarding the methodology of modal metaphysics. And in fact anyone who thinks we can do metaphysics in the present, a time before we know the fundamental theory true of the actual world, seems compelled to do the same.

The conclusion of this section can therefore be put as follows. If we think we can do metaphysics at all prior to the emergence of a fundamental theory then we should regard FOUNDATIONALISM as false. It is of course open to us to deny the antecedent; indeed, I have argued elsewhere that there are prima facie good reasons to do so (McKenzie 2020b). But since there is something of a performative contradiction in saying so much out loud in an essay on metaphysics, written in the here and now, my parting words are that FOUNDATIONALISM is false and no longer deserves inclusion as a component of the grounding orthodoxy.

5 Concluding Remarks

I opened this Element with a question: given that so many metaphysicians are physicalists, why do so few pay attention to physics when developing the metaphysics of fundamentality? I noted that the stock answer is that metaphysicians regard their work as conceptually prior to that of the physicist. As such, only once the metaphysician has defined what it is to be 'ontologically prior' or 'ontologically fundamental' can the physicist get to work and tell us what it is that fits the bill. That was not a position that seemed plausible to me. And I tried to back this up by showing how attending to the theories and practices of physics can shed new light on live debates in metaphysics – in particular, one regarding the nature of the grounding relation (that of whether it is 'unified'), and one regarding its structure (that of the status of FOUNDATIONALISM).

For all that, the conclusions I arrived at were not categorical. My argument against the unity of grounding was only compelling to committed Humeans (of which I am not one), and that against foundationalism could be resisted by biting some methodological bullets (albeit some rather large ones). For all that the arguments hewed closely to science, then, they landed in the typical predicament of metaphysics – a subject whose conclusions, as I noted above in Section 3, are characteristically less compulsory than those of science. But what I hope nevertheless to have made *unambiguous* is that the metaphysician who regards grounding as a largely proprietary concept, with physics relegated to providing the relevant inventory only once the hard metaphysical work is done, needlessly deprives themselves of theoretical resources that can both illuminate and go some way to resolve the questions that they set themselves.

References

Abell, C. (2012). 'Art: What It Is and Why It Matters'. In: *Philosophy and Phenomenological Research* 85.3, pp. 671–691. http://doi.org/10.1111/j.1933-1592.2011.00498.x.

Aristotle. (1984). *The Complete Works of Aristotle (Vol. 2, pp. 1552–1728)*. Ed. J. Barnes. Princeton University Press.

Armstrong, D. M. (1983). *What Is a Law of Nature?* Cambridge University Press.

— (1978). *Universals and Scientific Realism: A Theory of Universals (Vol. 2)*. Cambridge University Press.

Audi, P. (2012). 'Grounding: Toward a Theory of the In-Virtue-Of Relation'. In: *Journal of Philosophy* 109.12, pp. 685–711. http://doi.org/10.5840/jphil20121091232.

Barnes, E. (2018). 'Symmetric Dependence'. In: *Reality and Its Structure*. Ed. R. L. Bliss and G. Priest. Oxford University Press. pp. 50–69.

Baron, S. and Norton, J. (2019). 'Metaphysical Explanation: The Kitcher Picture'. In: *Erkenntnis*, 86.1, pp. 187–207. http://doi.org/10.1007/s10670-018-00101-2.

Beebee, H. (2000). 'The Non-Governing Conception of Laws of Nature'. In: *Philosophical and Phenomenological Research* 61.3, pp. 571–594. http://doi.org/10.2307/2653613.

Bennett, K. (2017). *Making Things Up*. Oxford University Press.

Berker, S. (2018). 'The Unity of Grounding'. In: *Mind* 127.507, pp. 729–777. http://doi.org/10.1093/mind/fzw069.

Bertulani, C. A. and Danielewicz, P. (2019). *Introduction to Nuclear Reactions*. CRC Press.

Bhogal, H. (forthcoming). 'Nomothetic Explanation and Humeanism About Laws of Nature'. In: *Oxford Studies in Metaphysics*. Oxford University Press.

Bigaj, T. and Wüthrich, C. (2016). 'Introduction'. In: *Metaphysics in Contemporary Physics*. Ed. T. Bigaj and C. Wüthrich. Brill Rodolpi, pp. 1–6.

Bitbol, M. (2007). 'Schrodinger against Particles and Quantum Jumps'. In: *Quantum Mechanics at the Crossroads*. Ed. J. Evans and A. S. Thorndike. Springer, pp. 81–106.

Bliss, R. and Priest, G. (2018). 'Introduction'. In: *Reality and Its Structure*. Ed. R. L. Bliss and G. Priest, Oxford University Press, pp. 1–33.

Bliss, R. and Trogdon, K. 2019: 'Metaphysical Grounding', In *The Stanford Encyclopedia of Philosophy (Winter 2019 Edition)*. Ed. Edward N. Zalta. https://plato.stanford.edu/archives/win2019/entries/grounding.

Bliss, R. L. (2013). 'Viciousness and the Structure of Reality'. In: *Philosophical Studies* 166.2, pp. 399–418. http://doi.org/10.1007/s11098-012-0043-0.

Bohm, D. (1957). *Causality and Chance in Modern Physics*. Routledge & Kegan Paul.

Bohn, E. D. (2018). 'Infinitely Descending Ground'. In: *Reality and Its Structure*. Ed. R. L. Bliss and G. Priest. Oxford University Press, pp. 167–181.

Born, M. (1956). 'Physics and Metaphysics'. In: *The Scientific Monthly* 82.5, pp. 229–235.

Brown, R. and Ladyman, J. (2009). 'Physicalism, Supervenience and the Fundamental Level'. In: *Philosophical Quarterly* 59.234, pp. 20–38. http://doi.org/10.1111/j.1467-9213.2008.613.x.

Bryant, A. (2018). 'Naturalizing Grounding: How Theories of Ground Can Engage Science'. In: *Philosophy Compass,* 13.5, e12489. http://doi.org/10.1111/phc3.12489.

Callender, C. (2001). 'Why Be a Fundamentalist: Reply to Schaffer'. http://philsci-archive.pitt.edu/archive/00000215/S.

Calosi, C. and Morganti, M. (2018). 'Interpreting Quantum Entanglement: Steps Towards Coherentist Quantum Mechanics'. In: *British Journal for the Philosophy of Science,* vol. 72, no. 3, p. axy064. http://doi.org/10.1093/bjps/axy064.

Calosi, C. (2020). 'Priority Monism, Dependence and Fundamentality'. In: Philosophical Studies 177.1, pp. 1–20.

Cameron, R. P. (2018). 'Infinite Regress Arguments'. In: *The Stanford Encyclopedia of Philosophy (Fall 2018 Edition)*. Ed. Edward N. Zalta. https://plato.stanford.edu/archives/fall2018/entries/infinite-regress.

Cameron, R. P. (2008). 'Turtles All the Way Down: Regress, Priority and Fundamentality'. In: *Philosophical Quarterly* 58.230, pp. 1–14. http://doi.org/10.1111/j.1467-9213.2007.509.x.

(2019). 'Truthmaking, Second-Order Quantification, and Ontological Commitment'. In: *Analytic Philosophy* 60.4, pp. 336–360. http://doi.org/10.1111/phib.12162.

Cappelli, A. et al. (2012). 'Introduction to Part II'. In: *The Birth of String Theory*. Ed. by A. Cappelli et al. Cambridge University Press, pp. 83–99.

Chakravartty, A. (2017). *Scientific Ontology: Integrating Naturalized Metaphysics and Voluntarist Epistemology*. Oxford University Press.

Chew, G. F. (1964). 'Elementary Particles'. In: *Physics Today* 17.4, pp. 30–33.

(1968). '"Bootstrap": A Scientific Idea?' In: *Science* 161, pp. 762–765.

Correia, F. (2021). 'A Kind Route from Grounding to Fundamentality'. In: *Synthese*, pp. 1–17. http://doi.org/10.1007/s11229-021-03163-y.

Craver, C. F. (2014). 'Levels'. In: *Open MIND* 8(T).9–10. Frankfurt am Main: MIND Group. http://doi.org/10.15502/9783958570498.

Dasgupta, S. (2014). 'The Possibility of Physicalism'. In: *Journal of Philosophy* 111.9–10, pp. 557-592. http://doi.org/10.5840/jphil20141119/1037.

  (2017). 'Constitutive Explanation'. In: *Philosophical Issues* 27.1, pp. 74–97. http://doi.org/10.1111/phis.12102.

Dorr, C. (2010). 'Review of James Ladyman and Don Ross, *Every Thing Must Go: Metaphysics Naturalized*'. In: *Notre Dame Philosophical Reviews* 2010.6.

Emery, N. (2019). 'Laws and Their Instances'. In: *Philosophical Studies* 176.6, pp. 1535–1561. http://doi.org/10.1007/s11098-018-1077-8.

Fine, K. (2001). 'The Question of Realism'. In: *Philosophers' Imprint* 1, pp. 1–30.

  (2012). 'Guide to Ground'. In: *Metaphysical Grounding*. Ed. F. Correia and B. Schnieder. Cambridge University Press, pp. 37–80.

French, S. (2014). *The Structure of the World: Metaphysics and Representation*. Oxford University Press.

French, S. and McKenzie, K. (2012). 'Thinking Outside the Toolbox: Towards a More Productive Engagement between Metaphysics and Philosophy of Physics'. In: *European Journal of Analytic Philosophy* 8.1, pp. 42–59.

  (2016). 'Rethinking Outside the Toolbox: Reflecting Again on the Relationship between Philosophy of Science and Metaphysics.' In: *Metaphysics in Contemporary Physics*. Ed. T. Bigaj and C. Wüthrich. Brill Rodopi, pp. 25–54.

Friedman, M. (2018). 'Ernst Cassirer'. In: *Stanford Encyclopedia of Philosophy*. Ed. Edward. N. Zalta. Fall 2018. Metaphysics Research Lab, Stanford University. https://plato.stanford.edu/archives/fall2018/entries/cassirer

Glazier, M. (2020). 'Explanation'. In: *Routledge Handbook of Metaphysical Grounding*. Ed. M. J. Raven, pp. 121–132. Routledge.

Guay, A. and Pradeu, T. (2020). 'Right Out of the Box: How to Situate Metaphysics of Science in Relation to Other Metaphysical Approaches'. In: *Synthese* 197.5, pp. 1847–1866. http://doi.org/10.1007/s11229-017-1576-8.

Hakkarainen, J. (2015). 'Hume on Spatial Properties'. In: *Nominalism About Properties: New Essays*. Ed. G. Ghislain and R. Gonzalo. Routledge, pp. 79–94.

Hall, N. 2010. David Lewis's Metaphysics. In *The Stanford Encyclopedia of Philosophy (Fall 2010 Edition)*. Ed. Edward N. Zalta. http://plato.stanford.edu/archives/fall2010/entries/lewis-metaphysics.

Hicks, M. T. and Schaffer, J. (2017). 'Derivative Properties in Fundamental Laws'. In: *British Journal for the Philosophy of Science* 68.2, pp. 411–450. http://doi.org/10.1093/bjps/axv039.

Hildebrand, T. (2020). 'Non-Humean Theories of Natural Necessity'. In: *Philosophy Compass* 15.5, e12662. http://doi.org/10.1111/phc3.12662.

Hoefer, C. (2003). 'For Fundamentalism'. In: *Philosophy of Science* 70.5, pp. 1401–1412. http://doi.org/10.1086/377417.

Hofweber, T. (2009). 'Ambitious, Yet Modest, Metaphysics'. In: *Metametaphysics: New Essays on the Foundations of Ontology*. Ed. D. J. Chalmers, D. Manley, and R. Wasserman. Oxford University Press, pp. 260–289.

Imaguire, G. (2020). 'Ontological Categories and the Transversality Requirement'. In: *Grazer Philosophische Studien* 97.4, pp. 619–639. http://doi.org/10.1163/18756735-000116.

Jaworski, W. (2011). *Philosophy of Mind: A Comprehensive Introduction*. Wiley-Blackwell.

Jeans, J. (1915). *The Mathematical Theory of Electricity and Magnetism*. Cambridge University Press.

Jenkins, C. (2011). 'Is Metaphysical Dependence Irreflexive?' In: *The Monist* 94.2, pp. 267–276. http://doi.org/10.5840/monist201194213.

Koslicki, K. (2015). 'The Coarse-Grainedness of Grounding'. In: *Oxford Studies in Metaphysics* 9, pp. 306–344.

Kovacs, D. M. (2019). 'Ricki Bliss and Graham Priest (Eds.): Reality and Its Structure: Essays in Fundamentality (Review)'. In: *Notre Dame Philosophical Reviews* 2019.17. https://urldefense.com/v3/__ https://ndpr.nd.edu/news/reality-and-its-structure-essays-in-fundamentality/__; !!Mih3wA!WP84psfxEr1VPg6Z-0Z-TXMOsORycJiN85ZWITu Nez-LCgkZf0njkjg4817SkuI$.

— (2021). 'The Oldest Solution to the Circularity Problem for Humeanism About the Laws of Nature'. In: *Synthese,* 198, pp. 8933–8953. http://doi.org/10.1007/s11229-020-02608-0.

Ladyman, J. and Ross, D. (2007). *Every Thing Must Go: Metaphysics Naturalized*. Oxford University Press.

Leuenberger, S. (2020). 'The Fundamental: Ungrounded or All-Grounding?' In: *Philosophical Studies* 177.9, pp. 2647–2669. http://doi.org/10.1007/s11098-019-01332-x.

Lewis, D. K. (1983). *Philosophical Papers*. Oxford University Press.

(1986). 'Against Structural Universals'. In: *Australasian Journal of Philosophy* 64.1, pp. 25–46. http://doi.org/10.1080/00048408612342211.

Lieb, E. (1990). 'The Stability of Matter: From Atoms to Stars'. In: *Bulletin of the American Mathematical Society* 22, pp. 1–49.

Lieb, E. H. and Loss, M. (2001). *Analysis (2nd ed.)*. Graduate Studies in Mathematics, Vol. 14. American Mathematical Society.

Litland, J. E. (2013). 'On Some Counterexamples to the Transitivity of Grounding'. In: *Essays in Philosophy* 14.1, p. 3. http://doi.org/10.7710/1526-0569.1453.

Loewer, B. (2007). 'Laws and Natural Properties'. In: *Philosophical Topics* 35.1/2, pp. 313–328. http://doi.org/10.5840/philtopics2007351/214.

(2012). 'Two Accounts of Laws and Time'. In: *Philosophical Studies* 160.1, pp. 115–137. http://doi.org/10.1007/s11098-012-9911-x.

Loux, M. J. (2006). *Metaphysics: A Contemporary Introduction (3rd ed.)*. Routledge.

Lowe, E. J. (1998). *The Possibility of Metaphysics: Substance, Identity, and Time*. Clarendon Press.

MacBride, F. and Janssen-Lauret, F. (2015). 'Meta-Ontology, Epistemology & Essence: On the Empirical Deduction of the Categories'. In: *The Monist* 98.3, pp. 290–302. http://doi.org/10.1093/monist/onv015.

Maudlin, T. (2007). *The Metaphysics within Physics*. Oxford University Press.

Maurin, A.-S. (2007). 'Infinite Regress - Virtue or Vice?' In: *Hommage À Wlodek*. Ed. T. Rønnow-Rasmussen et al. http://www.fil.lu.se/HommageaWlodek.

Maurin, A.-S. (2019). 'Grounding and Metaphysical Explanation: It's Complicated'. In: *Philosophical Studies* 176.6, pp. 1573–1594.

McKenzie, K. (2011). 'Arguing against Fundamentality'. In: *Studies in History and Philosophy of Science Part A* 42.4, pp. 244–255. http://doi.org/10.1016/j.shpsb.2011.09.002.

(2017a). 'Against Brute Fundamentalism'. In: *Dialectica* 71.2, pp. 231–261. http://doi.org/10.1111/1746-8361.12189.

(2017b). 'Relativities of Fundamentality'. In: *Studies in History and Philosophy of Science Part B: Studies in History and Philosophy of Modern Physics* 59, pp. 89–99.

(2020a). 'A Curse on Both Houses: Naturalistic versus a Priori Metaphysics and the Problem of Progress'. In: *Res Philosophica* 97.1, pp. 1–29. http://doi.org/10.11612/resphil.1868.

(2020b). 'Structuralism in the Idiom of Determination'. In: *British Journal for the Philosophy of Science* 71.2, pp. 497–522. http://doi.org/10.1093/bjps/axx061.

McKenzie, K. and Muller, F. (2017). 'Bound States and the Special Composition Question'. In: *EPSA15 Selected Papers*. Ed. Michela Massimi, Jan-Willem Romeijn, and Gerhard Schurz. Springer, pp. 233–241.

McSweeney, M. M. (2020). 'Debunking Logical Ground: Distinguishing Metaphysics from Semantics'. In: *Journal of the American Philosophical Association* 6.2, pp. 156–170. http://doi.org/10.1017/apa.2019.40.

Mehta, N. (2017). 'Can Grounding Characterize Fundamentality?' In: *Analysis* 77.1, pp. 74–79. http://doi.org/10.1093/analys/anx044.

Melnyk, A. (2003). *A Physicalist Manifesto: Thoroughly Modern Materialism*. Cambridge University Press.

Moore, A. W. (2011). *The Evolution of Modern Metaphysics: Making Sense of Things*. Cambridge University Press.

Morganti, M. (2015). 'Dependence, Justification and Explanation: Must Reality Be Well-Founded?' In: *Erkenntnis* 80.3, pp. 555–572. http://doi.org/10.1007/s10670-014-9655-4.

Ney, A. (2020). 'Are the Questions of Metaphysics More Fundamental Than Those of Science?' In: *Philosophy and Phenomenological Research* 100.3, pp. 695–715. http://doi.org/10.1111/phpr.12571.

Nolan, D. (2014). 'Hyperintensional Metaphysics'. In: *Philosophical Studies* 171.1, pp. 149–160. http://doi.org/10.1007/s11098-013-0251-2.

North, J. (2013). 'The Structure of a Quantum World'. In: *The Wave Function: Essays on the Metaphysics of Quantum Mechanics*. Ed. A. Ney and D. Albert. Oxford University Press, pp. 184–202.

Norton, J. D. (2014). 'A Material Dissolution of the Problem of Induction'. In: *Synthese* 191.4, pp. 1–20. http://doi.org/10.1007/s11229-013-0356-3.

Passmore, J. (1961). *Philosophical Reasoning*. Gerald Duck.

Paul, L. A. (2012a). 'Metaphysics as Modeling: The Handmaiden's Tale'. In: *Philosophical Studies* 160.1, pp. 1–29. http://doi.org/10.1007/s11098-012-9906-7.

(2012b). 'Building the World from its Fundamental Constituents'. In: *Philosophical Studies* 158.2, pp. 221–256. http://doi.org/10.1007/s11098-012-9885-8.

Popper, K. R. (1962). *Conjectures and Refutations: The Growth of Scientific Knowledge*. Basic Books.

Rabin, G. O. (2018). 'Grounding Orthodoxy and the Layered Conception'. In: *Reality and Its Structure*. Ed. R. L. Bliss and G. Priest. Oxford University Press, pp. 37–50.

Rappoccio, S. (2019). 'The Experimental Status of Direct Searches for Exotic Physics Beyond the Standard Model at the Large Hadron Collider'. In: *Reviews in Physics* 4, p. 100027. issn: 2405-4283.

Raven, M. J. (2015). 'Ground'. In: *Philosophy Compass* 10.5, pp. 322–333. http://doi.org/10.1111/phc3.12220.

— (2017). 'New Work for a Theory of Ground'. In: *Inquiry: An Interdisciplinary Journal of Philosophy* 60.6, pp. 625–655. http://doi.org/10.1080/0020174X.2016.1251333.

Rescher, N. (1983). 'The Unpredictability of Future Science'. In: *Physics, Philosophy, and Psychoanalysis*. Ed. R. Cohen and L. Laudan. D. Reidel, pp. 153–168.

Reutlinger, A. and Saatsi, J. (2018). *Explanation Beyond Causation: Philosophical Perspectives on Non-Causal Explanations*. Oxford University Press.

Rodriguez-Pereyra, G. (2005). 'Why Truthmakers?' In: *Truthmakers: The Contemporary Debate*. Ed. H. Beebee and J. Dodd. Oxford University Press, pp. 17–31.

Rosen, G. (2010). 'Metaphysical Dependence: Grounding and Reduction'. In: *Modality: Metaphysics, Logic, and Epistemology*. Ed. B. Hale and A. Hoffmann. Oxford University Press, pp. 109–136.

Saatsi, J. (2017). 'Explanation and Explanationism in Science and Metaphysics'. In: *Metaphysics and the Philosophy of Science: New Essays*. Ed. M. Slater and Z. Yudell. Oxford University Press, pp. 162–191.

Schaffer, J. (2003). 'Is There a Fundamental Level?' In: *Noûs* 37.3, pp. 498–517.

— (2008). 'Causation and Laws of Nature : Reductionism'. In: *Contemporary Debates in Metaphysics*. Ed. T. Sider, J. Hawthorne, and D. W. Zimmerman. Blackwell, pp. 82–107.

— (2009). 'On What Grounds What'. In: *Metametaphysics: New Essays on the Foundations of Ontology*. Ed. y D. Manley, D. J. Chalmers, and R. Wasserman. Oxford University Press, pp. 347–383.

— (2010). 'Monism: The Priority of the Whole'. In: *Philosophical Review* 119.1, pp. 31–76. http://doi.org/10.1215/00318108-2009-025.

— (2012). 'Grounding, Transitivity, and Contrastivity'. In: *Metaphysical Grounding: Understanding the Structure of Reality*. Ed. F. Correia and B. Schnieder. Cambridge University Press, pp. 122–138.

(2016). 'Grounding in the Image of Causation'. In: *Philosophical Studies* 173.1, pp. 49–100. http://doi.org/10.1007/s11098-014-0438-1.

Seager, W. (2016). *Theories of Consciousness: An Introduction and Assessment*. Routledge.

Sider, T. (2011). *Writing the Book of the World*. Oxford University Press.

(2013). 'Symposium on Writing the Book of the World'. In: *Analysis* 73.4, pp. 751–770. http://doi.org/10.1093/analys/ant085.

(2020). 'Ground Grounded'. In: *Philosophical Studies* 177.3, pp. 747–767. http://doi.org/10.1007/s11098-018-1204-6.

Skiles, A. (2015). 'Against Grounding Necessitarianism'. In: *Erkenntnis* 80.4, pp. 717–751. http://doi.org/10.1007/s10670-014-9669-y.

Thompson, N. (2016). 'Grounding and Metaphysical Explanation'. In: *Proceedings of the Aristotelian Society* 116.3, pp. 395–402.

Veltman, M. J. (2003). *Facts and Mysteries in Elementary Particle Physics*. World Scientific.

Werner, J. (2020). 'A Grounding-Based Measure of Relative Fundamentality'. In: *Synthese,* 198.10, pp. 9721–9737. http://doi.org/10.1007/s11229-020-02676-2.

Westerhoff, J. (2018). 'Metaphysical Vertigo'. In: *Studies in the Ontology of E. J. Lowe*. Ed. T. Tambassi. Editiones Scholasticae, Neunkirchen-Seelscheid, pp. 27–46.

Williams, N. E. (2019). *The Powers Metaphysic*. Oxford University Press.

Wilsch, T. (2015). 'The Nomological Account of Ground'. In: *Philosophical Studies* 172.12, pp. 3293–3312. http://doi.org/10.1007/s11098-015-0470-9.

Wilson, A. (2018a). 'Metaphysical Causation'. In: *Noûs* 52.4, pp. 723–751. http://doi.org/10.1111/ nous.12190.

Wilson, J. M. (2014). 'No Work for a Theory of Grounding'. In: *Inquiry: An Interdisciplinary Journal of Philosophy* 57.5–6, pp. 535–579. http://doi.org/10.1080/0020174x.2014.907542.

(2018b). 'Grounding-Based Formulations of Physicalism'. In: *Topoi* 37.3, pp. 495–512. http://doi.org/10.1007/s11245-016-9435-7.

Acknowledgements

This is a short book but it got a lot of help. I'd like to thank Craig Callender, Eddy Keming Chen, Eugene Chua, Sam Elgin, Anncy Thresher, and Porter Williams for helpful feedback early in the planning process. Steven French and F.A. Muller deserve special thanks for reading and commenting on the whole thing, as do Jacob Stegenga and two incisive but anonymous referees. Thanks go also to Nic Fillion and Fred Muller again for helping me with the diagrams, and to David Meyer for walking me through a crucial proof. Reuven Brandt deserves a big thanks for the tech support and moral support that I've come to take for granted.

This was written during the 2020 pandemic and would not have been possible without everyone who helped to take care of my kids. As such, special thanks go to Lucy and Abe, Alexis and Clover, Linda and Shayla, Cathy, Cami, Mari, and Nara.

Cambridge Elements ≡

Philosophy of Science

Jacob Stegenga
University of Cambridge

Jacob Stegenga is a reader in the Department of History and Philosophy of Science at the University of Cambridge. He has published widely on fundamental topics in reasoning and rationality and philosophical problems in medicine and biology. Prior to joining Cambridge he taught in the United States and Canada, and he received his PhD from the University of California San Diego.

About the Series

This series of Elements in Philosophy of Science provides an extensive overview of the themes, topics and debates which constitute the philosophy of science. Distinguished specialists provide an up-to-date summary of the results of current research on their topics, as well as offering their own take on those topics and drawing original conclusions.

Cambridge Elements ≡

Philosophy of Science

Elements in the Series

Scientific Knowledge and the Deep Past: History Matters
Adrian Currie

Philosophy of Probability and Statistical Modelling
Mauricio Suárez

Relativism in the Philosophy of Science
Martin Kusch

Unity of Science
Tuomas E. Tahko

Big Data
Wolfgang Pietsch

Objectivity in Science
Stephen John

Causation
Luke Fenton-Glynn

Philosophy of Psychiatry
Jonathan Y. Tsou

Duhem and Holism
Milena Ivanova

Bayesianism and Scientific Reasoning
Jonah N. Schupbach

Fundamentality and Grounding
Kerry McKenzie

A full series listing is available at: www.cambridge.org/EPSC

Printed in the United States
by Baker & Taylor Publisher Services